D0432022

SPACE ODYSSEY

Voyaging through the Cosmos

Dust, debris, and reflected light tint the far reaches of the universe.

SPACE ODYSSEY
Voyaging through the Cosmos

WILLIAM HARWOOD

NATIONAL GEOGRAPHIC

WASHINGTON, D.C.

Meteor trail slashes the night sky above California's Joshua Tree National Park.

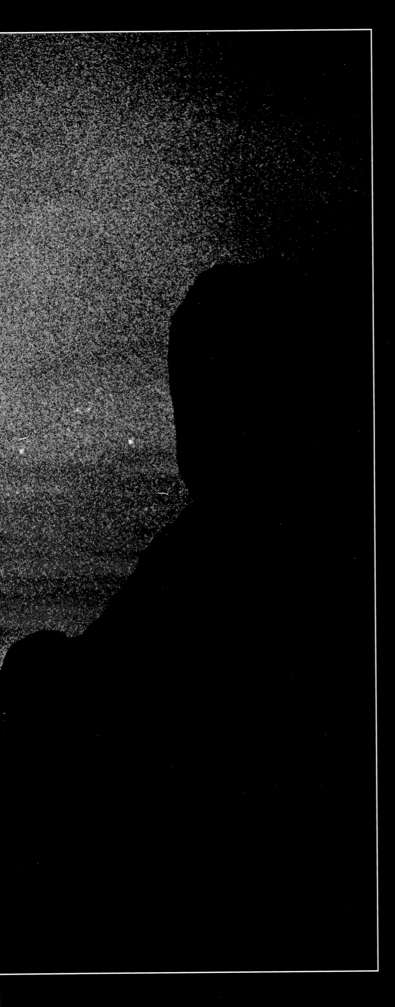

CONTENTS

INTRODUCTION

The biggest single problem with space exploration is...space! It's virtually impossible for the human mind to grasp the enormity of a single galaxy, let alone the entire universe.

It's one thing to know that a beam of light traveling 186,000 miles every second—fast enough to circle the world 7.5 times in a single heartbeat—would take 30,000 years to reach the center of our galaxy. It's another matter entirely to comprehend what a six-trillion-mile-long light-year is in the first place. It's hard enough to visualize the scale of our own solar system; at 7.3 billion miles across, it's so vast that sunlight takes five and a half hours to reach Pluto.

To put such numbers in perspective, imagine Earth, currently home to six billion human beings, as a single grain of sand floating over home plate in a professional baseball stadium. At that scale, writes Edward Packard in *Imagining the Universe*, the Sun would be the size of a golf ball ten feet behind the batter. Mars would be a speck of dust a weak bunt away; Jupiter, no bigger than an apple seed, would be found about three-quarters of the way to the pitcher's mound. Saturn would be 90 feet out, near first base; Uranus and Neptune would be pinheads floating in left field, one shallow, one deep; and Pluto would be a dust mote near the center field wall some 400 feet from the batter's box.

With our solar system compressed to the size of a baseball field in San Francisco, the nearest star, Proxima Centuri, would be shining nearly 600 miles away, a single peppercorn somewhere in the Grand Canyon. Sirius, the brightest star in the sky, would be a tennis ball blazing away somewhere near Topeka, Kansas.

Now imagine launching a spacecraft from that grain of sand back at home plate in San Francisco. Imagine further that our spacecraft travels at the scaled-down equivalent of 25,000 mph, the peak speed of an Apollo moonship and the fastest any human has ever moved with respect to Earth. At that velocity, it would take our intrepid astronaut more than 16 years to reach Pluto in deep center field—and some 118,000 years to reach the nearest star. At interstate highway speeds,

the jaunt from San Francisco to the Grand Canyon would take 42 million years!

Now let's change the scale again. Imagine that the grain of sand atop home plate in San Francisco is not Earth but the entire solar system. On this scale, Packard writes, the Milky Way would span 13 miles across; the Andromeda galaxy, the most distant object visible to the unaided eye, would be another 13-mile-wide spiral 300 miles away, in Nevada. Not too hard to visualize. But even then—reducing our solar system to the size of a single grain of sand—the ever-receding edge of the observable universe would lie more than two million miles away.

The key problem for would-be star travelers is the speed limit that's woven into the fabric of the universe: 186,000 miles per second. The central equations in Einstein's relativity theory require division by zero for anything to achieve the speed of light, and division by the square root of a negative number to exceed it. Neither operation is possible, of course, and until a new theory emerges that either supplants relativity or absorbs it in some broader framework, faster-than-light travel will remain the stuff of science fiction—and stars will remain points of light twinkling in the darkness.

But the very equations that forbid faster-than-light travel offer a loophole of sorts. For as a spacecraft approaches the speed of light, the onboard passage of time as viewed by an outside observer slows down—an effect known as time dilation. Astronauts aboard a ship traveling at a constant 99.9999 percent of the speed of light—185,981 miles per second—would appear to age less than one month in the course of covering a distance of four light-years. But while such relativistic

Segmented primary mirror and gangly tracker of the domed Hobby-Eberly telescope stand ready atop Mount Fowlkes in Texas. The telescope achieved full operation in late 1998; its mission—spectroscopy, or the study of individual wavelengths of light—can reveal the temperature, chemical composition, and motion of astronomical objects.

KNOWN UNIVERSE

Like soap bubbles adrift in air, the 125 billion or so galaxies of the visible universe are separated by huge voids. While gravity binds together stars of a given galaxy and the galaxies of a given cluster, all clusters and groups are racing away from each other due to the Big Bang, a titanic explosion believed to have taken place some 14 billion years ago, spawning the enormity of the universe from a single point in space-time.

LOCAL SUPER-CLUSTER

This great aggregation of clusters of separate galaxies spans more than a hundred million light-years. Centered on the Virgo cluster, it includes Ursa Major and other clusters, each of which contains thousands of separate galaxies. Total mass of our supercluster is roughly equivalent to that of a quadrillion— a thousand trillion— Suns. Yet most of the space it occupies is totally empty.

LOCAL GROUP

A relatively small cluster on the outskirts of our supercluster, our local galaxy group extends over three million light-years from the Milky Way, our home galaxy, and includes two other large spiral galaxies, Andromeda (M31) and Triangulum (M33). It is receding from Virgo as the universe continues to expand.

Andromeda Galaxy

Triangulum Galaxy

Small Magellanic Cloud

Large Magellanic Cloud

Milky Way

space travel is possible in theory, for all practical purposes it is no more achievable than the faster-than-light "warp drive" invoked by *Star Trek*'s writers.

Still, who can say what might be possible in the next 50 or 100 years? Within the past 70 years, humans have gone from the Wright Flier to Apollo 11; we've landed men on the Moon and sent robotic spacecraft to every planet in the solar system except Pluto. Four of those spacecraft are now headed for interstellar space, the first crude starships known to history.

As the first decade of the 21st century unfolds, a new era in space exploration is underway. The United States, European nations, and Japan are dispatching an armada of sophisticated spacecraft to the distant corners of the solar system. Powerful new ground- and space-based telescopes are being devised to probe the far reaches of the universe. Closer to home, astronauts are living in orbit aboard the International Space Station, while others contemplate a return to the Moon and eventual flights to Mars.

This ongoing space odyssey promises nothing less than a revolution in our understanding of the universe and Earth's place in it. By the middle of the next decade, scientists may find out whether life ever

MILKY WAY GALAXY

About 100,000 light-years in diameter, our home galaxy contains a few hundred billion stars, concentrated in a bulging core and flat, spiral arms. Just as planets circle the Sun, the Milky Way's member stars orbit the galactic core, believed to be centered on a gigantic black hole.

A MATTER OF PERSPECTIVE

Truly colossal in scale, the universe we can see spans roughly 28 billion light-years, or about 165 sextillion (that's 165,000,000,000,000,000,000,000) miles. How to fathom the unfathomable? By splitting it into smaller and smaller subunits, each of which includes our home star and home planet.

LOCAL STARS

In the Orion arm of the Milky Way, our local stellar neighborhood includes stars within 20 to 30 light-years of Earth. Most are too dim to be seen without telescopes, but a few, such as Sirius, are familiar beacons in the night sky. The nearest star to us, Proxima Centauri, is a red dwarf possessing only a tenth of our Sun's mass, lying some 4.3 light-years away.

OUTER SOLAR SYSTEM

The outermost planets of our solar system include the four "gas giants"—huge, cold, gaseous orbs with rings and numerous moons—and the icy dwarf we call Pluto. Beyond lies the flat Kuiper Belt and the far larger, spherical Oort cloud, both of which harbor comets. All the planets, comets, and the Sun are believed to have condensed about 4.6 billion years ago from the disklike solar nebula, a great cloud of dust and gas.

INNER SOLAR SYSTEM

Four rocky "terrestrial planets" orbit nearest the Sun, accompanied by thousands of much smaller rocky asteroids, most of which occupy the asteroid belt between Jupiter and Mars. Earth, the only known abode of life, is outstanding also for its surface aggregations of liquid water as well as its volcanism, mountain building, seafloor spreading, continental drift, and other dynamic processes.

Sirius

Sun

Alpha Centauri

Pluto

Neptune

Uranus

Saturn

Jupiter

Sun

Mars

Earth

Venus

Mercury

Sun

Asteroid Belt

evolved on Mars or even on Jupiter's icy moon, Europa. They will learn whether Earth-like planets are common or rare and, possibly, whether any of those newly discovered worlds harbor life.

And who knows? Maybe someone will figure out a way around Einstein's speed limit. So fasten your seat belts and get ready for a most fantastic voyage.

Earth's home galaxy, the Milky Way spreads a cluster of dark dust, hundreds of billions of stars, and glowing red gases across our evening sky. At lower left, Earth's own clouds reflect the day's last moments of sunlight.

PART I

What's Out There

THE SOLAR SYSTEM IN PERSPECTIVE

O n February 14, 1990, NASA's aging Voyager 1 probe, looking back on the solar system from a point 3.7 billion miles out from the Sun, obediently snapped a final few dozen pictures as it sailed into the star-sprinkled night of interstellar space. Thirty-nine of those grainy snapshots, assembled into a remarkable mosaic, show six of the solar system's planets—Venus, Earth, Jupiter, Saturn, Uranus, and Neptune—in the one and only such family portrait ever made.

Tiny Mercury was lost in the glare of the Sun, Mars was never positively identified, and remote Pluto—smaller than our Moon and no longer thought to be a planet in the traditional sense of the word—was too dim for Voyager's camera to distinguish. In close-ups of the six visible planets, only Jupiter and Saturn show discernible disks. Earth appears as a tiny pinprick of reflected sunlight offering no hint

Above a Canadian lakeshore, the northern lights dance over a Dene Indian tepee, both timeless symbols of the wild. Earth's auroras—borealis in the Northern Hemisphere, australis in the Southern— flicker to life when charged solar particles collide with molecules in our atmosphere, causing them to emit photons of energy—the colored lights that we see.

that it is home to more than six billion thinking beings and the only known biosphere in the solar system—or anywhere else, for that matter.

It is a photograph that forces the viewer to contemplate the scale of the solar system, the vast distances between its planets and the even greater gulfs that separate stars in the larger galaxy beyond. It drives home the realization that Earth is a mote adrift in a sea of space, that humanity, if not alone, is at the very least isolated on a truly grand scale. The tiny gap on the photomosaic that separates the Sun from the light speck that is Earth in reality yawns 93 million miles, what astronomers refer to as one astronomical unit. Neptune is 30 times farther removed; Pluto orbits, on average, another 9.5 astronomical units beyond that.

Think about it like this: A rifle bullet fired from the Sun at 3,000 feet per second and—for the sake of argument—immune to gravity and other external forces, would need more than 200 years to reach Pluto's average distance. Even light, traveling at an unimaginable 186,000 miles per second, needs five and a half hours to make the trip. It is no wonder that humanity's home in space seems a tiny point of light from Voyager 1. The real wonder is that it can be seen at all.

The solar system photographed by Voyager 1 is, one would assume, typical of solar systems around other stars in the Milky Way: Small, rocky, Earth-like planets basking in the warmth of the Sun, while larger, volatile-rich gas giants similar to Jupiter and Saturn orbit in the colder reaches farther out. A tenuous disk

Eyes in the sky—orbiting satellites of the Defense Meteorological Satellite Program—reveal a glittering Earth. The eastern United States, Europe, and Japan glow most brightly, while much of the southern hemisphere remains dark. Scientists use such data to map the effects of urbanization; they currently estimate that, worldwide, movement toward cities is growing at three times the rate of population expansion. Or, as one says, "like mold on an orange."

of small, frozen planetesimals spread too thinly to have had a shot at becoming planets extends deep into space beyond the orbit of Neptune. Beyond that is a vast cloud of unseen comets, extending a full light-year or more in all directions.

The Sun, of course, is the beating heart that holds this whirling complex of planets, moons, asteroids, and comets together, its nuclear furnace providing the energy that, among other things, sustains life on Earth. Its gravitational grip extends more than a light-year into space; the solar wind, a high-speed stream of charged particles blasting out from the Sun's surface in all directions, defines a bubble in space more than 200 astronomical units across.

How did our solar system form, and why did it evolve in this particular manner? Did it follow some universal blueprint for solar system construction, or is Earth's system somehow unique? As of June 18, 2001, 58 confirmed extrasolar planets—that is, planets not of our solar system—had been discovered orbiting other stars, while five more were circling collapsed pulsars. By the time you read this, those numbers will have grown as astronomers around the world engage in friendly but heated competition to discover as many extrasolar planets as current technology allows.

Solar systems discovered to date show unexpected variety. In some, huge gas-giant planets many times the mass of Jupiter orbit surprisingly near their parent stars, completing a "year" in just a few days. Is this blueprint more common than our solar system's? Or is it simply an artifact of the current techniques used to detect extrasolar planets, which happen to favor discovering Jupiter-class giants close to their parent suns? The answer could shed light on how common—across our home galaxy—life as we know it might be.

"We've found Saturn- and Jupiter-size objects, but not out where Saturn and Jupiter are in our own solar system. They're in much closer," says Charles Beichman, senior program scientist with NASA's long-range effort to find extrasolar planets. "And they've found them around 7 percent, roughly, of the stars like our Sun. In those 7 percent, the orbital dynamics are probably such that no Earth could survive with these giant planets screeching around. Imagine a freeway or a giant parking lot with these big buses careening around and you're sitting in your little Volkswagen trying to make it from one end to the other. You're going to get clobbered. So these are probably systems where Earths and life are very difficult to come by."

By the end of the second decade of the 21st century, U.S. and European spacecraft now on the drawing board will be able to directly detect Earth-size planets around other stars, allowing astronomers to resolve the uniqueness question once and for all. But for now, the only solar system astronomers know in any detail is Earth's, made up of eight major planets, countless smaller bodies like Pluto, 91 known moons, thousands of asteroids, and billions of primordial comets.

Mercury, Venus, Earth, and Mars, in order from the Sun out, make up the four inner, rocky, "terrestrial" planets while Jupiter, Saturn, Uranus, and Neptune are the outer "gas giants." Between Mars and Jupiter, a broad belt of asteroids composed of rocky debris remains from the birth of the solar system. Jupiter's titanic gravity presumably prevented this material from forming a planet. Even if it had, the resulting body would have been less massive than our Moon.

Pluto, long considered the solar system's ninth and outermost planet, now appears to be one of the larger representatives of a vast disk of more than 200 million asteroid-size bodies making up what is known as the Kuiper belt. This tenuous disk is believed to extend in a broad ring from Neptune's orbit out to a distance of 50 to 150 astronomical units.

Kuiper belt objects, perhaps disturbed by gravitational interactions with undiscovered Pluto-class bodies even farther out, occasionally plunge into the inner solar system to become short-period comets, that is, comets that complete a trip around the Sun in less than 200 years. Halley's comet, which completes an orbit every 76 years, is an example. Pluto and even Neptune's strange moon, Triton, may be Kuiper belt objects that were displaced aeons ago by one or more catastrophic collisions.

Is the outer region of the Kuiper belt, then, the boundary of the solar system? Not yet! Beyond lies the Oort cloud, a spherical shell made up of billions if not trillions of small, cold chunks of dirty ice. Their combined mass is close to Earth's. This huge shell extends up to three light-years from the Sun. Oort cloud objects apparently originated near the gas giants early in the history of the solar system. Many were ejected into interstellar space by gravitational encounters with Jupiter, Saturn, Uranus, and Neptune.

Gravitational interactions with nearby stars or the outer planets also can cause a member of the Oort cloud to fall toward the Sun, becoming a so-called long-period comet in the process. Such comets typically return to the Oort cloud or exit the solar system entirely. Examples include comets Hale-Bopp and Hyakutake. Long-period comets pose a particularly nasty threat to Earth because they cannot be seen until they are within the orbit of Jupiter. By then, there would be little time to do anything should they be on a collision course with Earth.

Details about the origin and evolution of this menagerie of planets, moons, asteroids, and comets are not fully understood. But astronomers believe they have a fairly good idea of the general outline. Picture yourself in this general area of the Milky Way galaxy 4.6 billion years ago. You'd probably see a vast nebula extending hundreds of light-years in all directions, with infant stars glowing like hot embers inside shrouds of swirling gas. This nebula consists primarily of molecular hydrogen and helium, with trace amounts of oxygen, carbon, neon, nitrogen, magnesium, silicon, iron, and a sprinkling of other heavy elements. The hydrogen and helium were created in the Big Bang that gave birth to the universe. Almost everything else was cooked up by countless generations of stars.

You can see a similar cloud today. Find the constellation Orion and look for the middle star in the hunter's sword. Binoculars show that "star," some 1,500 light-years away, is actually a softly glowing nebula, a hazy, cloudlike structure. A small telescope reveals delicate wisps and fans extending like the wings of some great bird. The Hubble Space Telescope, operating above Earth's obscuring atmosphere, sees here a stellar nursery where dozens of infant stars and planetary systems are in the process of being born.

Areas of slightly higher density in such primordial clouds, called cloud cores, eventually can accumulate enough mass to overcome magnetic fields that tend to resist gravitational collapse. As more and more material falls inward, the central regions of these cloud cores heat up and begin glowing: They become protostars. But not all the infalling material reaches the central core. Some is diverted through complex interactions into a flat, rotating disk.

"Once the accumulating disk achieves a mass roughly one-third that of the protosun, it becomes gravitationally unstable and changes from an elegant, symmetric item of cosmic dinnerware to something less regular, perhaps resembling a miniature galaxy," writes geologist John Wood, an expert in planetary

evolution, in *The New Solar System*. "It may be that star-disk systems grow by a series of excursions into instability of this sort."

For ten million years or so, the central star continues growing until the temperatures in its core reach levels high enough to initiate self-sustaining nuclear fusion reactions, in which hydrogen nuclei combine to produce heavier helium nuclei, releasing enormous energy in the process. And so a star is born.

"During this time, the associated dusty disk becomes less and less evident," Wood writes. "Some of its substance probably continues to migrate into the central star; much of the residual gas may be heated so greatly by the star's ultraviolet radiation that it evaporates into interstellar space." A tiny fraction of the dust and gas that pass through the disk may aggregate, he adds, "into discrete lumps that remain in orbit around the star after the disk is gone, serving as the seeds of its planetary family."

The nature and chemical composition of those lumps depends on where they form in the solar nebula that gives rise to the protoplanetary disk. Temperatures are highest near the star, dropping off lower and lower farther away, and defining three general zones.

In the innermost zone of Earth's developing solar system, it was too warm for water to condense as ice; planetary lumps here consisted of rocky silicates and other elements that vaporize only at high temperatures. They evolved into the terrestrial planets: Mercury, Venus, Earth, and Mars.

"The next zone was colder, water ice was stable, and a vast blizzard of snowflakes gave rise to the much larger Jovian planets," Wood continues. "In the outermost and thus coldest zone, condensed matter was also icy. But it was too sparsely distributed to accrete into sizable planets; instead it remained dispersed in small icy planetesimals—comets—in what we now call the Kuiper belt." Remarkably, he adds, "the planets assembled themselves very quickly. Although the process differed in detail from zone to zone, virtually everything was in place within ten million years, by which time the solar nebula had largely dissipated."

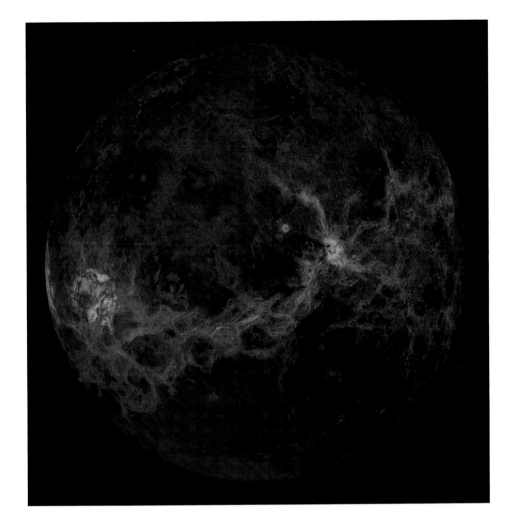

Named for the goddess of love and beauty, **Venus shines more brightly in our sky than anything except the Sun and the Moon. The planets of our solar system range in size from massive Jupiter—more than 300 times the size of Earth—to minuscule Pluto, which some astronomers now argue might be demoted from planet status to mere "Kuiper belt object." All the planets orbit the Sun in the same direction and lie in roughly the same plane. Scientists believe they were formed some 4.6 billion years ago.**

Questions still remain, of course, such as how the planets and their moons acquired their various tilts and rotations. Pluto, Venus, and Uranus spin backward compared to neighboring planets, while the rotational axes of most planets are tilted at varying angles with respect to the Sun's equatorial plane. These variations might be explained by collisions with extremely large bodies early in the solar system's history.

"This is consistent with accretion not from a large number of small bodies, but from a smaller number of large bodies, some of which were *quite* large," Wood writes. "In this case, the sum of all the planetesimals' contributions would have been unequal, leaving residual spins and tilts." Saturn's relatively large tilt, for example, likely required an oblique impact with an object more massive than all the terrestrial planets combined.

The dozens of moons that now whirl about the gas giants, making these large planets seem like mini-solar systems, present another puzzle. Some of them undoubtedly formed in place as the initially hot planets cooled and contracted. Others, like Neptune's Triton and the two moons of Mars, Phobos and Deimos, probably were captured later during passing encounters. But Earth's Moon, it now seems, was the result of a catastrophic collision between our home planet and a large, fast-moving planetesimal that blasted molten debris into space. Some of that material fell back to Earth, some escaped into interplanetary space, and some settled into a disk around the planet. The disk material eventually coalesced into the large Moon we enjoy today.

Astronomers once debated whether our solar system formed as the result of a single catastrophic event or as the result of natural evolutionary processes. In the former, solar systems would be relatively rare; in the latter, they would be common. It now appears that elements of both contribute to the solar system we see around us and, by extension, those that formed around other stars. Whether the layout of our solar system is a common design or an unusual departure from the norm is not yet known. But with any luck at all, the answers are just around the corner.

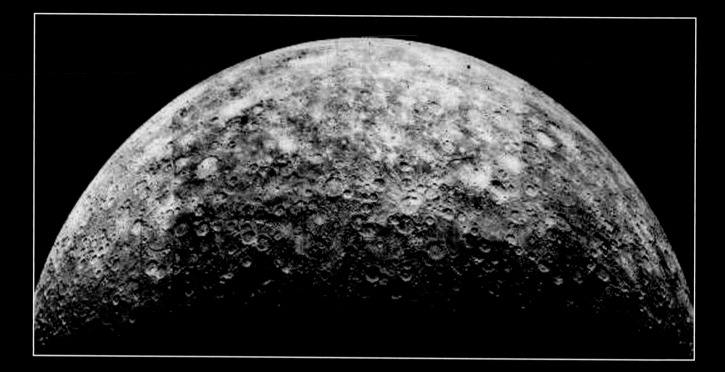

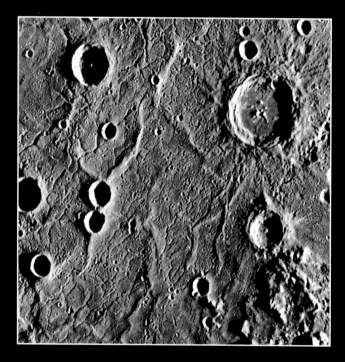

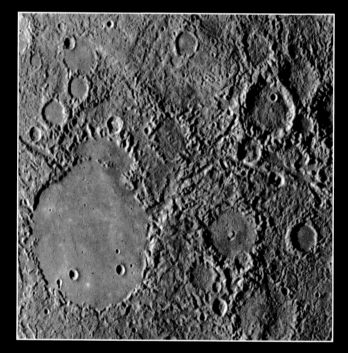

MERCURY More pocked than our Moon, Mercury (top) lies nearest the Sun. Its largest feature, the enormous Caloris Basin (above, right) was born of a cataclysmic impact about 3.8 billion years ago; shock waves rippled through the planet, creating a region of linear depressions and jumbled hills on Mercury's opposite side (above, left). Unlike other planets, Mercury cannot be studied through the Hubble Space Telescope because its nearness to the blazing Sun would damage Hubble's delicate instruments.

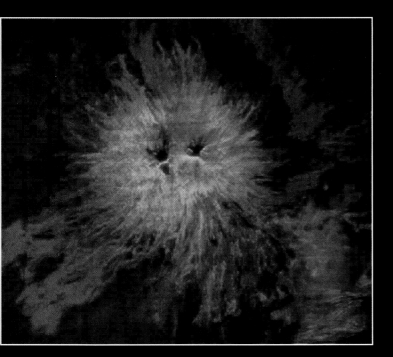

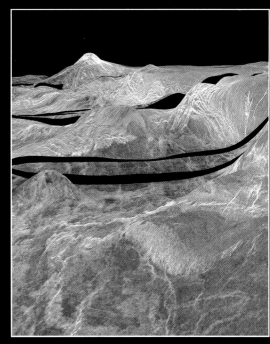

VENUS False-color image of Venus (top) combines years of data; colors represent different altitudes. Mile-high volcano Sapas Mons (above, left) resembles Earthly hot-spot volcanoes such as those in Hawaii. The area shown here is about 400 miles across. Three-dimensional perspective (above, right) captures the southern scarp-and-basin province of western Ishtar Terra. On Venus, our nearest planetary neighbor, each day is 243 Earth days long—long enough to make a Venusian day slightly longer than a Venusian year.

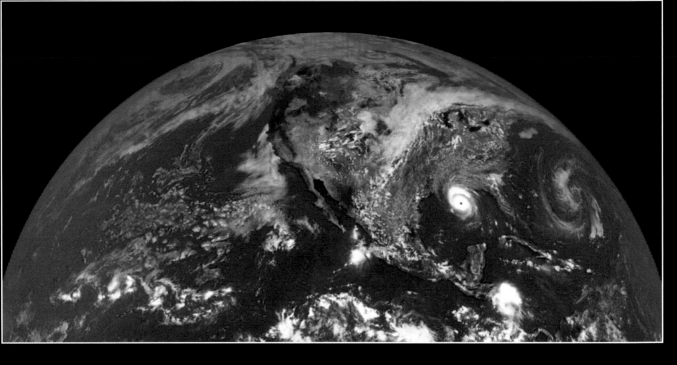

EARTH Until Copernicus in the 16th century, the water planet (top) was considered the center of the universe. Its plasmasphere—a blanket of ionized gas—was long suspected but never seen until a robot spacecraft took this picture (right) in 2000, from high above the North Pole. An all-sky map (below) helps scientists judge Earth's relative speed through the universe: Radiation in the direction of motion appears blueshifted, while radiation on the opposite side is redshifted. It clocks us hurtling through space at about 373 miles a second.

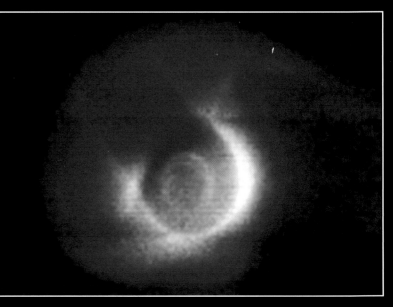

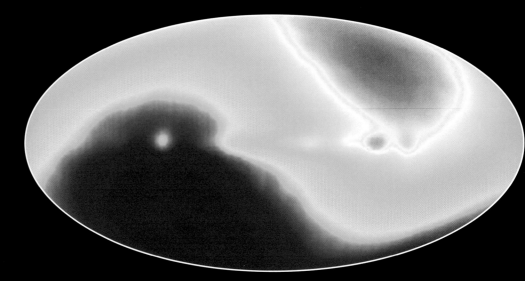

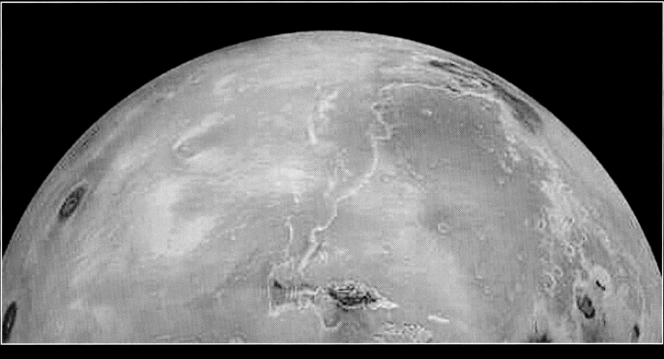

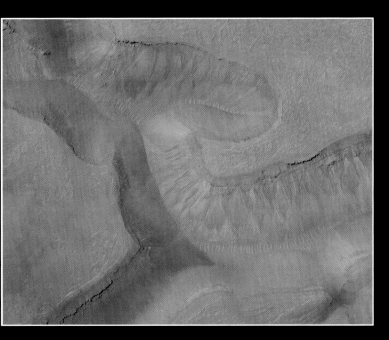

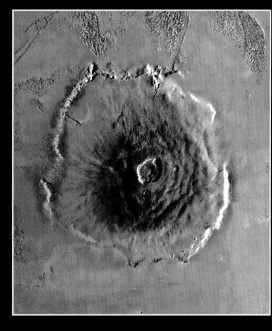

MARS Favorite of science-fiction writers, Mars (top)
is named for the Roman god of war. Polar ice caps
and signs of erosion lead scientists to believe there
may be liquid water on Mars—one of the prerequi-
sites for life. Data from MOLA—the Mars Orbiter
Laser Altimeter—produce high-resolution topo-
graphic shade maps of the red planet (above, left).
The instrument collects about 900,000 measurements
of elevation every day. Olympus Mons (above, right),
the largest volcano on Mars, is 400 miles across and
climbs 17 miles into the thin Martian atmosphere.

THE GAS GIANTS

Systems of their Own

"Our ignorance about planet formation is strongly dependent on distance from the Sun," reflects Alan Boss, a leading planetary theorist with the Carnegie Institution in Washington, D.C. "We kind of know how the terrestrial planets formed; everyone pretty much agrees they formed by banging together rocky bodies. But when you get out to Jupiter and Saturn, there are two competing theories. When you get out to Uranus and Neptune, there are probably four or five. And none of them is really great."

Welcome to the outer solar system, a truly alien realm dominated by four giant, distinctly different planets that hold sway over complex mini-solar systems made up of at least 87 distinctly different moons. Two of these moons are larger than Mercury, four are larger than Earth's Moon, one has a thick primordial

Jupiter, one of the four giant outer planets with huge dense atmospheres surrounding small rocky cores, shows nine different faces to the Cassini Orbiter during one revolution in November 2000. Reddish orange bands indicate different zones of clouds; some flow eastward, some to the west. Some frames reveal the Great Red Spot, an enormous storm permanently ensconced in the planet's southern hemisphere.

atmosphere, and at least two others may harbor vast subsurface oceans.

All four gas giants have globe-circling ring systems made up of ice and fractured bits of rock herded along in lockstep by the gravity of shepherding moons. All four feature powerful magnetic fields and windy, volatile-rich atmospheres, some with long-lived storms big enough to swallow Earth. Whether the gas giants have solid cores or not is unknown, but three—Jupiter, Saturn, and Neptune—generate more heat through gravitational contraction and friction than they receive from the distant Sun.

Tiny Pluto, traditionally the Sun's ninth and outermost planet, is a mere mote compared to the gas giants. It is not only smaller than the other planets but also smaller than seven of the solar system's moons. Currently viewed as a comet-like Kuiper belt object rather than a true planet, it has little in common with the giants of the outer solar system beyond a striking similarity to Neptune's largest moon, Triton.

Jupiter, of course, is the solar system's largest planet, a gargantuan world some 318 times more massive than Earth. It contains 71 percent of the solar system's total planetary mass. Saturn has 95 times the mass of our planet, Uranus 14.5, and Neptune 17.1. Altogether, the gas giants account for 99.6 percent of our solar system's planetary mass and all but four of its 91 known moons. The four terrestrial planets—Mercury, Venus, Earth, and Mars—together contain slightly less than just two Earth masses of material.

But mass is just one distinguishing characteristic of the gas giants. The diameters of Jupiter and Saturn are, respectively, 11.2 and 9.4 times greater than Earth's. If they were dinner plates, Earth would be roughly the size of a dime. Uranus and Neptune are each about 4 times wider than Earth. All four gas giants are much less dense than the rocky inner planets, and Saturn is even less dense than water; if you could find an ocean big enough to hold it, Saturn would float.

Nearly a hundred times more massive than Earth, Saturn has been known to mankind since prehistoric times. Galileo first observed it by telescope in 1610, but his primitive instrument could not resolve the rings. A half century later, the Dutch astronomer Christiaan Huygens discovered their true nature. Composed of countless small particles, the essentially flat rings are just a few hundred feet thick but span 155,000 miles or more in diameter.

As for moons, Jupiter has at least 28, second only to Saturn's current total of 30. Uranus has 21, while Neptune has 8 and possibly more. All four gas giants have been visited by at least one spacecraft for close-up observations. NASA's Pioneer 10 flew past Jupiter in 1973; Pioneer 11 visited both Jupiter and Saturn, in 1974 and 1979, respectively. Both Voyager probes also flew past Jupiter and Saturn, and Voyager 2 continued on to Uranus for a 1986 flyby before zipping past Neptune and icy Triton in 1989. Jupiter was visited a fifth time, in 1992, when the Ulysses solar probe, a joint project of NASA and the European Space Agency, used the planet's gravity to fling it into orbit around the Sun's poles. As it flew by, Ulysses studied Jupiter's powerful magnetic field in unprecedented detail.

NASA's sophisticated Galileo spacecraft, the first designed to orbit an outer planet for long-term observations, reached Jupiter in late 1995 and dropped an instrumented probe into its atmosphere before beginning an ever-changing orbital tour permitting repeated flybys of major Jovian moons. Now nearly out of fuel, Galileo is focusing on the volcanic moon Io, before going out in a blaze of glory in October 2003 with a kamikaze plunge into Jupiter's atmosphere.

"The Jupiter guys have a very interesting set of data now," says Torrence Johnson, Galileo project scientist at the Jet Propulsion Laboratory in Pasadena. "It's not like it overturned all our ideas about Jupiter. But we now kind of know which of the various competing ideas after Voyager are on the right track."

NASA's Cassini spacecraft, launched in 1997, is currently heading for Saturn to make long-term studies of that planet's atmosphere, its glorious ring system, its cloud-shrouded moon, Titan, and other enigmatic satellites. In 1999, Cassini became the seventh space-craft to fly past Jupiter, making coordinated observations with the Galileo craft.

The realm of the gas giants extends roughly 2.3 billion miles—from Jupiter, orbiting the Sun at an average distance of 484 million miles, to Neptune, an average 2.8 billion miles out. It's an enormous span, some 25 times as far as Earth is from the Sun.

The terrestrial planets are rocky because they coalesced in the inner, warmer regions of the solar nebula's disk, where temperatures were high enough to essentially boil off lighter, more volatile elements. The gas giants, says Boss, followed a somewhat differ-ent script: "The conventional wisdom is you make the cores of these planets—if they have cores—by banging together solid bodies. Only not just rocky bodies, icy bodies, too. So you build up a core and then you have to pull gas in from the disk, which makes the bulk of the [planet's] mass." Scientists believe Jupiter's core probably is equivalent in mass to six Earths, blanketed by another 312 Earth masses of gas.

While this core accretion theory represents the conventional view, "it has a problem in that it's a slowpoke way of making Jupiter," Boss adds. "It takes about eight million years to do it. And most (proto-planetary) disks don't last that long. In most cases, those disks are gone in a few million years, in which case you wouldn't end up with a Jupiter. So that tells you either planets like Jupiter must be rare, or else maybe there's another way of making Jupiter faster."

One possibility, the disk instability theory, holds that small instabilities in the cooler, outer regions of a protoplanetary disk cause material to clump together "in one fell swoop, producing a self-gravitating Jupiter-mass object of gas and dust," Boss says. "Then the dust grains inside that object coagulate and settle down towards the center and form the core afterward." But, he adds, this technique won't work much past the orbit of Neptune, because there is simply not enough material present in the protoplanetary disk. "There's sort of a magic Goldilocks zone, around 5 to 10 to 20 astronomical units, where things are massive enough and cold enough that you can make the clumps grow."

Both scenarios maintain that Jupiter and Saturn grew by sucking in hydrogen, helium, and other con-stituents from the original solar nebula. Data from the Galileo spacecraft, however, show that Jupiter today has less helium and neon, and more argon, krypton, and xenon, than expected when compared to the Sun. Jupiter's lack of helium and neon can be explained by a theory that helium migrates toward a planet's high-pressure interior; in effect, it rains liquid helium. But Jupiter's relatively high levels of the other gases represent a different puzzle.

"At least one major fairy story that's going around is that the only way to get those enhancements of the rare gases [argon and xenon] in the atmosphere is to have the material that was brought into Jupiter be con-densed at very low temperatures where the ices will trap those gases," says Johnson. "None of the models of the early solar nebula allow those temperatures to be that low at Jupiter's distance. So either a lot of material is being brought in from much colder areas, or Jupiter started out forming in colder areas and migrated to its current position."

In recent years, Jupiter-class planets have been discovered orbiting their parent stars at extremely close range, within a few million miles. Galileo's findings tend to support the idea that giant planets can migrate from their original orbits, due either to gravitational interaction with the protoplanetary disk or to encoun-ters with other nearby planets.

Wherever Jupiter and Saturn first took shape, both grew large and fast enough to gravitationally draw in material directly from the protoplanetary disk. The terrestrial planets did not have the gravitational muscle for such feats. Neither, perhaps, did Uranus or Neptune. While the core accretion theory remains the generally accepted explanation for those two planets, some theorists believe this would take too long, given the tenuous nature of the outer regions of the protoplanetary nebula and the infrequency of the collisions necessary to build them up. Calculations show, Boss says, that the disk would have thinned out and dissipated long before Uranus and Neptune could have started gravitationally attracting material as Jupiter and Saturn apparently did. Disk instability might have played a role in the evolution of these outermost gas giants, but then some other mechanism must have acted later, to strip off the outer layers. Interestingly, Uranus rotates on its side, roughly in the plane of the solar system, indicating this planet may have endured a catastrophic collision with another massive body in the remote past that knocked it off-kilter. Such a collision would have stripped away mass, but probably not enough to explain the planet's current size.

However they originated, all four gas giants are as fascinating as they are diverse, none more so than Jupiter, with its great size and titanic gravity, its family of intriguing moons, its delicate rings, and its turbulent atmosphere. By chance, the probe that Galileo dropped into Jupiter's atmosphere hit a relatively dry region marked by huge downdrafts. About 60 miles below Jupiter's uppermost clouds of ammonia ice, it found a hazy, generally cloud-free sky with a pressure of 1.6 atmospheres. Researchers had expected a much cloudier, hotter, and more humid environment, with much higher pressures.

As it turned out, the water *was* there, just not where the probe entered. Additional data revealed huge storms at the interfaces between Jupiter's differentially rotating horizontal cloud belts, phenomena that may be duplicated on Saturn, if not Uranus and Neptune.

"We think we know where the water is now; we think we know where the thunderstorms are," Johnson says. "They're in the large cyclonic-belt zone boundaries.

They're very, very highly concentrated in these areas. They're the stormy zones on Jupiter. We had hints of that from Voyager, but not everybody believed it." Data from Galileo also show that high-altitude jet streams do not die off with depth in the atmosphere, suggesting that internal heating from gravitational contraction plays a greater role in powering a gas giant's atmosphere than does energy from the Sun.

In another major finding, Galileo photographed the innermost quartet of Jupiter's many moons and showed the two farthest out—Amalthea and Thebe—are directly responsible for the planet's three major rings. "We kind of hit the jackpot with that," Johnson reflects. "Presumably all the rings are effectively the product of material that's been kicked off of these small satellites by micrometeoroid impact. That was a very clear signature."

Four of Jupiter's moons—Io, Europa, Ganymede, and Callisto—are all easily visible from Earth with binoculars or a small telescope. Discovered by Italian astronomer Galileo Galilei in 1610 and known collectively as the Galilean satellites, they hold perhaps the biggest surprises of all. Certainly they are the most visually dramatic, running the gamut from fire to ice.

Consider Io. With a diameter of 2,263 miles, Io is slightly larger than Europa, Pluto, or our Moon. But it orbits Jupiter just 262,000 miles from the center of the planet. As such, it is alternately squeezed and stretched by Jupiter's powerful gravity, producing tidal heating worthy of Dante's inferno.

Geysers of sulfur dioxide jet skyward and lava continuously resurfaces this entire world. If you could stand on Io, says Johnson, "you'd see flows of lava, some of them cooling off, some of them actively being refreshed. There's enough lava coming out of Io every year to cover the whole surface to an average depth of a centimeter [0.4 inch] or more. It would be in a sense like standing on a whole world that was just out there in the middle of one of the eruptions in Hawaii. Every place on the satellite is like that."

Ganymede also has yielded a major surprise: It possesses a magnetic field, the first ever discovered

in a moon, and a feature that is difficult to explain. Magnetic fields are thought to originate in electrically conducting liquid layers in a planet's interior. But according to current models, Ganymede's core should no longer be hot enough to provide the necessary convection.

Data from Galileo strongly support earlier hints that Europa may have a layer of liquid water just below its frozen crust. This finding has encouraged NASA to consider sending an orbiter to map Europa's subsurface world; another mission could penetrate the ice and explore the ocean firsthand. And Europa may not be alone; Callisto also might harbor a subsurface sea. Although this moon does not have a core as such, magnetometer data suggest the presence of an electrically conducting layer of some sort—which might be explained by a saltwater ocean. Spectral studies also indicate the presence of sulfur and carbon dioxide, which may be trapped in surface ice. Scientists also see evidence of carbon-nitrogen bonds like those in complex organic molecules.

Like a galactic Old Faithful, the volcano Prometheus regularly spews dark plumes of gas and dust across Io, one of Jupiter's numerous moons. The lighter brown semicircle near the top is a lava-filled caldera. Many active volcanoes constantly rework Io's surface; they result from Jupiter's relentless gravitational pull, which powers enormous tidal stresses and internal friction. Tidal flexing causes Io's surface to move up and down more than 300 feet a day.

"Taken together, the data provide the first direct evidence that icy satellites contain the carbon, nitrogen, and sulfur compounds common in primitive meteorites and comets," Johnson writes in *Scientific American*. "These materials are also some of the most important for life."

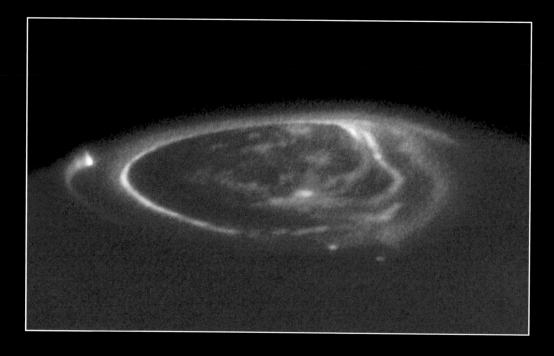

JUPITER Possibly sighted by telescope as early as the 17th century, the planet's bulging Great Red Spot (above) has persisted at least since 1830. Winds raging along each of Jupiter's colorful bands—easterlies near the equator and westerlies at the poles—sustain it and other huge storms, or eddies. Two or three times wider than Earth, the spot easily dwarfs the moon Io, passing by at left. An electric-blue aurora (opposite, top) circles the planet's north magnetic pole, overlain with magnetic "footprints" left by three of Jupiter's largest natural satellites: Io at left, Ganymede near the center, and Europa. Fresh lava (opposite, bottom) erupts from a giant volcanic caldera on Io in February 2000. Multiple images taken through various filters aboard NASA's Galileo spacecraft were combined to create this color mosaic.

SATURN From 5.5 million miles out, Voyager 1 creates an enhanced-color image of Saturn's northern hemisphere (above). Storms rage in both the dark and light brown belts; a longitudinal wave runs the length of the light blue region. The smallest features are about 100 miles across. An assembly of Voyager 1 images (opposite, top) neatly portrays some of the at least 30 moons found so far in the Saturnian system: Dione in the foreground, Tethys and Mimas to the right, Enceladus and Rhea to the left, and Titan in the distance. In a computer rendering (opposite, bottom) two of Rhea's many craters—Izanagi, the larger, and Izanami—overlap, while a small meteor strikes the surface of Izanagi as Saturn sits on the horizon. Voyager 1 visited Saturn and some of its larger moons in November 1980.

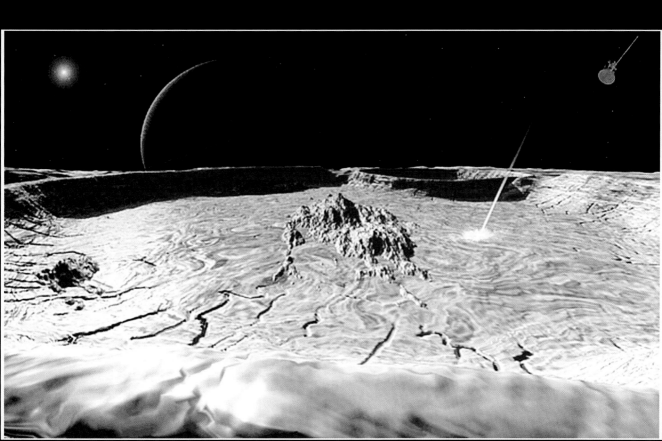

URANUS The seventh planet out from the Sun, Uranus has a mass 14.5 times Earth's. This false-color version (right), photographed by Voyager 2, actually combines three images taken with different filters to expose high-altitude hazes, including a pink area at the south pole. Uranus's rather featureless atmosphere appears blue-green naturally. The planet's nine main rings (below) occur here as horizontal lines. Fainter pastel lines between the rings result from computer enhancement.

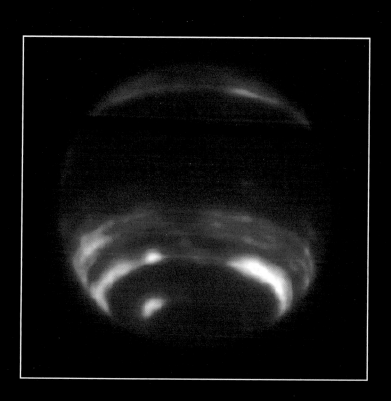

NEPTUNE Outermost gas giant, Neptune claims the solar system's fastest winds: Some blow at 1,250 miles an hour. Clouds (below) also circle the planet. The 10-meter Keck II telescope in Hawaii reveals Neptune in reflected sunlight (left). Bright bands represent layers of haze in the planet's upper atmosphere; a slit used for spectrographic analysis of light reflected from the planet shows up as a dark stripe.

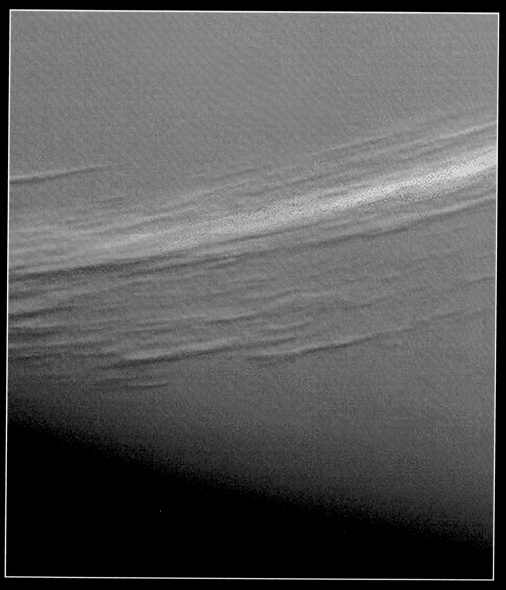

Visitors from the Depths of Space

Comets, Asteroids, and Interplanetary Dust

I n grammar school, children are taught that the solar system consists of the Sun, nine planets, their moons, a belt of asteroids between Mars and Jupiter, and a swarm of comets—dirty snowballs with gossamer tails that periodically grace the evening sky. Historically, all these bodies have been defined and treated as separate classes of objects: A planet is always a planet, a moon is a moon, and asteroids are distinct from comets.

Except when they're not. Pluto, long known as the ninth planet, may not be one after all. It is tiny—smaller than seven of the solar system's moons, including Earth's—and its orbit is unusually lopsided. Also, the small "moons" of Mars and the outer planets now seem to be captured asteroids, or perhaps dead comets that were plucked out of open space by chance gravitational encounters after the planets

Doomed to a fiery demise, a sungrazer comet—the thin white line near the bottom—hurtles toward our home star in this image from the Solar and Heliospheric Observatory (SOHO), which blocks out the Sun with a protective disk. Sungrazers apparently originated from a single large parent comet that broke up about 2,000 years ago. Approaching within 30,000 miles of the Sun, few survive the encounter.

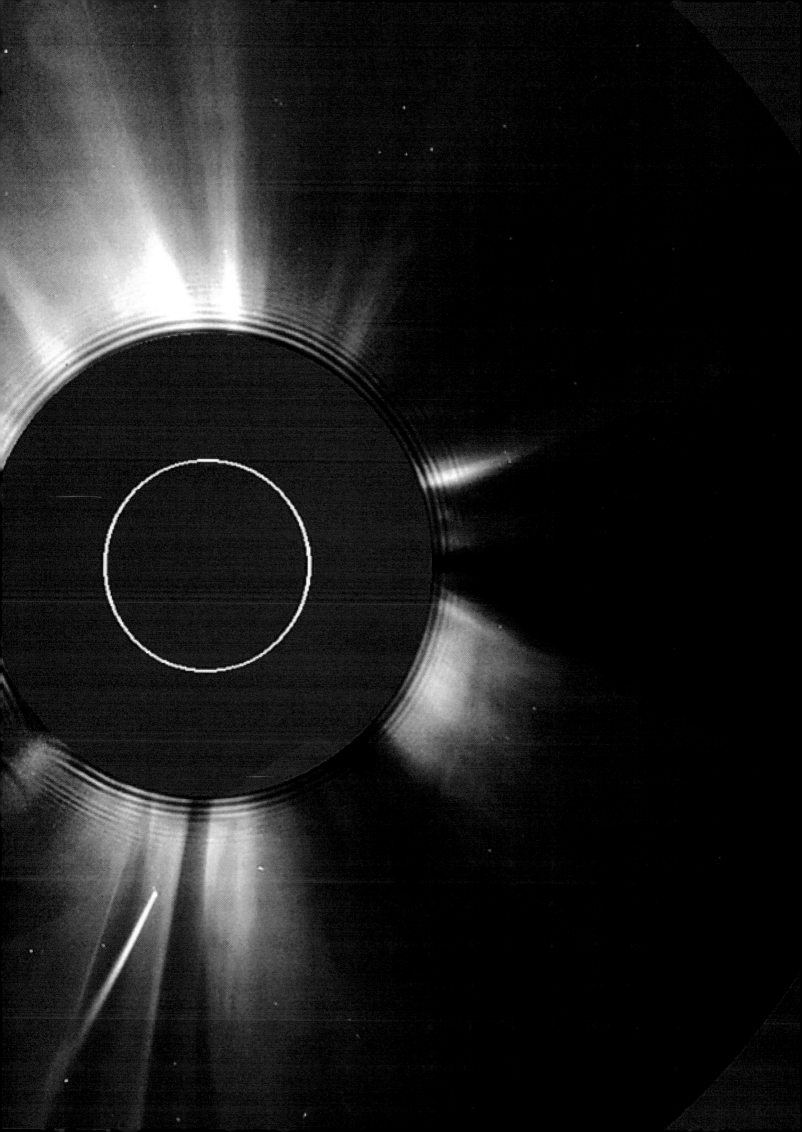

coalesced from the original solar nebula. Even the distinction between comets and asteroids is blurring as a new picture of the solar system emerges based on data from far-flung spacecraft, more sensitive ground instruments, and supercomputers that model how the Sun's family of planets, moons, asteroids, and comets has interacted over billions of years.

William Hartmann, a senior scientist at the Planetary Science Institute, believes astronomers in the past mistakenly, if understandably, viewed the solar system in a way that he calls the "nine-planets gestalt." In this perspective, the planets were viewed as the main items of interest and "all other bodies were looked upon as less important and less interesting."

"This hierarchical ordering—and particularly the disinterest accorded small objects—paved the way toward misleading distinctions among comets, asteroids, and certain moons," he writes in *The New Solar System*. "Now we are experiencing the breakdown of this gestalt. The Sun's family is seen instead as a complex system of worlds, 25 of which exceed 1,000 kilometers (621 miles) in diameter. Pluto is looking more and more like merely the largest member of the Kuiper belt. One might accurately describe our planetary system as eight major planets and a host of smaller worlds that were trapped in specific locations during the planet-forming process."

During that process 4.6 billion years ago, the temperature gradient across the solar nebula played a defining role in the chemical structure and compositional stratigraphy of the resulting solar system. The four terrestrial planets, as we've seen, formed close to the Sun where its warmth baked out hydrogen and other volatile elements. Jupiter and Saturn, in contrast, consisted mostly of hydrogen and helium collected directly from the primordial solar nebula, while Uranus and Neptune accreted volatile ices that condensed in the outer solar system.

So far, so good. But what of the comets and asteroids? Where did they originate, and what roles have they played in the solar system's evolution? While the planets may be the headliners in this cosmic drama, the smaller bodies, it now seems, at least deserve to share top billing.

"In terms of their relationship with life, they may well have brought the water and carbon-based molecules to the early Earth that allowed life to form," says Donald Yeomans, a researcher at the Jet Propulsion Laboratory who tracks comets and asteroids. "Subsequent collisions may have punctuated the evolution, wiped out the dinosaurs 65 million years ago and so allowed the mammals to move forward. So in a sense, we may owe our position atop the world's food chain to the fact that the dinosaurs checked out as a result of an impact. So I think there's an increased realization that comets and asteroids are not just the flotsam and jetsam of the solar system. They really are, next to the Sun itself, probably the most important objects in terms of power over life."

This realization has resulted in increased observation, theoretical analysis, and development of spacecraft to explore asteroids and comets first-hand. In the most stunning such flight to date, NASA put a $223-million spacecraft in orbit around asteroid Eros in 2000 for a full year of close-up observations. After taking more than 160,000 photographs and 11 million laser measurements, the Near Earth Asteroid Rendezvous mission—NEAR—ended in February 2001 with a dramatic touchdown on the surface of Eros.

"The principal reason we're looking at asteroids is because asteroids, meteorites, and comets are windows into the early solar system," Jessica Sunshine, a senior staff scientist with Science Applications International Corporation, said at a pre-landing news conference. "We want to understand what's going on in the early solar system because these fundamental materials and processes are what ultimately formed our planet and what we evolved from."

By the late 1990s, more than 8,000 asteroids had been detected, most in a thick doughnut-shaped torus between Mars and Jupiter. The largest of them is Ceres, discovered in 1801 and measuring about 580 miles across. Ceres alone represents about one quarter of all the mass in the main asteroid belt. The next biggest

Faint glow of Gegenschein—German for "counter glow"—lightens the night sky diametrically opposite the Sun. It occurs when asteroidal dust particles no larger than a millimeter, orbiting in the plane of the planets, reflect sunlight to Earth. During daylight hours, a similar phenomenon called "the glory" sometimes can be seen in clouds from airplanes.

asteroids, Pallas and Vesta, each span about 310 miles. By some estimates, there could be more than 500,000 with diameters of at least one mile. Even so, the total mass of all known asteroids is less than that of our Moon. And they're not all alike. About 75 percent are dark and carbonaceous (C-type). Another 15 percent are silicaceous, or stony (S-type); the final 10 percent consist of metallic (M-type) asteroids. Some appear to be made up of primordial material that is chemically unchanged since the birth of the solar system; others have undergone melting and differentiation much like that seen in terrestrial planets. Why this is so remains a major mystery.

In general, astronomers now believe that the planetesimals that formed between the orbits of Mars and Jupiter started out just like those that coalesced into planets, their chemical composition defined by their distance from the Sun. Their location also defined a transition zone between the inner and outer solar system, the so-called "frost line" beyond which volatiles condensed, giving rise to gas giants, icy moons, and billions if not trillions of comets. But gravitational disturbances disrupted the orbits of these planetesimals, preventing them from slowly accreting into a planet. Instead, they began colliding and breaking up. And the culprit was Jupiter.

"In order for objects to agglomerate, they have to run into each other slowly," says Yeomans. "Jupiter stirred them up so they didn't run into each other slowly, they were banging into each other."

In one scenario, large planetesimals near Jupiter were nudged by the planet's growing gravity into orbits that took them into what is now the asteroid belt. As they moved in and out of the belt, their own gravitational pull disrupted the orbits of smaller planetesimals, setting up collisions that continue even today. In another scenario, Jupiter's gravity directly affected the asteroid-belt planetesimals. Some were pumped up to high velocities and ejected from the belt, falling into the Sun or hitting other planets. Others were ejected from the solar system entirely. This process has continued throughout the solar system's history.

Two clumps of asteroids, one grouped ahead of Jupiter in its orbit and one just behind it, are known as Trojans. Another group, the Centaurs, occur much farther out, between the orbits of Jupiter and Neptune. More than 500 asteroids are known to approach or cross Earth's orbit. Those that come within 1.3 astronomical units of the Sun but do not cross Earth's orbit are known as Amor asteroids. Those that actually cross Earth's orbit are called Apollo asteroids, while those with orbits of less than one astronomical unit are called Aten asteroids. Astronomers believe more than 2,000 Aten-Apollo asteroids may be larger than 0.62 miles across, the threshold for causing global destruction.

All near-Earth asteroids pose a potentially significant threat to Earth. Some 65 million years ago, as Yeomans points out, an asteroid or comet measuring six miles or so across crashed into our planet near the tip of the Yucatan Peninsula and dug out a crater more than 100 miles wide. The resulting environmental catastrophe led to the extinction of the dinosaurs.

More than 150 other impact craters have been found on Earth, the largest nearly 200 miles across. A widespread search of the heavens is now underway to identify bodies large and near enough to pose a global threat to our planet.

In 1950, astronomer Fred Whipple proposed that comets were "dirty snowballs," or icy conglomerates. In the far reaches of the solar system, they're barely distinguishable from asteroids. But as they approach the Sun, their surface ice begins to vaporize, producing jets, long dust-and-plasma tails, and a dense, spherical envelope of gas and dust called a coma. During peak heating, comet Hale-Bopp lost 1,000 metric tons of dust and 130 tons of water every second. Even so, it probably loses less than one percent of its total mass every time it ventures into the inner solar system.

Comets come in two basic varieties: Long-period comets, like Hyakutake and Hale-Bopp, enter the inner solar system from random directions and take more than 200 years to complete one trip around the Sun; short-period comets take less than 200 years, and their orbits are tilted less than 30 degrees or so from the plane of the planets.

Dutch astronomer Jan Oort theorized that comets forming in the region of the outer planets eventually are thrown into the solar system's far reaches by gravitational interactions with the gas giants. As a result, he believed a huge cloud of comets should exist, extending a light-year or more in all directions. To explain the number of comets seen in the inner solar system, Oort calculated that the cloud that now bears his name must contain nearly 200 billion comets; current estimates put the number at several trillion. Passing stars and even giant molecular clouds can disturb the Oort cloud, at times shunting comets into the inner solar system. Most eventually return but others, like Halley's, are diverted into short-period orbits by gravitational encounters with one or more planets.

As it turns out, there is a second reservoir of comets in our solar system, predicted five decades ago by Irish theorist Kenneth Edgeworth and American planetary scientist Gerard Kuiper. Unlike the Oort

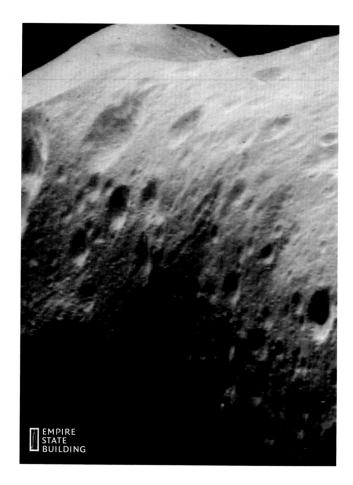

EMPIRE STATE BUILDING

cloud, which is generally spherical, this broad belt of comets—the Kuiper belt—is a relatively flat disk in the plane of the planets, extending from Neptune's orbit out to about 150 astronomical units. David Jewitt and Jane Luu discovered the first Kuiper belt object in 1992, a 200-mile-wide snowball in a 292-year orbit with an average distance of 44 astronomical units from the Sun. As of June 1, 2001, nearly 400 Kuiper belt objects had been found. The belt may contain 70,000 objects larger than 100 kilometers [62 miles], between 30 and 50 astronomical units out. Two of its more famous members are believed to be Pluto, discovered by Clyde Tombaugh in 1930, and Neptune's moon, Triton, discovered by William Lassell in 1846, just a few weeks after Johann Galle and Heinrich d'Arrest discovered Neptune itself.

Pluto measures just 1,430 miles wide; Triton is slightly larger, 1,680 miles in diameter. Their densities and surface temperatures are comparable, and both have surfaces made up of frozen methane, nitrogen, carbon monoxide, and water ice, implying that they had similar origins in the Kuiper belt.

Still, there are major differences. Triton circles Neptune in a retrograde (backward) orbit, while Pluto

Massive denizen of the asteroid belt, asteroid Eros (opposite) stretches about 20 miles in length. NASA's NEAR Shoemaker spacecraft made this image from 127 miles. An even closer view—only 22 miles away (above)—shows craters in different states of preservation: The one on the left retains a well-defined rim, while the older one at right is puckered by smaller craters and partly buried by fine debris. NEAR made a soft landing on Eros on February 12, 2001, after transmitting 69 close-up images during its descent.

orbits the Sun in its own lopsided orbit, accompanied by a relatively large moon, 737-mile-wide Charon, discovered by James Christy in 1978. Charon completes one orbit in the same amount of time it takes Pluto to complete one revolution about its axis: 6.4 days.

"If hypothetical Plutonians happened to live on the hemisphere directed away from Charon, they wouldn't even know Pluto had a moon," astronomer Dale Cruikshank writes in *The New Solar System*. "Conversely, for inhabitants of the Charon-facing side, the moon would always be in the sky at the same place, unchanging its location over the decades."

It appears likely Triton formed in the Kuiper belt and later collided with one of Neptune's original moons, destroying that moon and causing Triton's eccentric, retrograde orbit. Neptune's gravity and subsequent encounters with other moons eventually reshaped Triton's orbit into a more circular path. Astronomers are not sure how Pluto ended up with a moon in an orbit tilted 17 degrees to the plane of the planets. One scenario holds that Pluto formed in the Kuiper belt, with a nearly circular orbit. Charon was created later, after a violent collision pushed Pluto into a different orbit.

"However they came to be, we suspect that Triton and Pluto have something of a shared past," Cruikshank writes. "On the basis of their physical similarities and their proximity to the Kuiper belt, Triton and the Pluto-Charon binary are ever more frequently being regarded as very large members of the Kuiper belt. Pluto's status as a planet is thus challenged."

But not yet overthrown. Known for 70 years as the ninth planet from the Sun, Pluto's planetary ranking likely will remain in place for many years—if not in the minds of astronomers, then at least in the collective mind of the public.

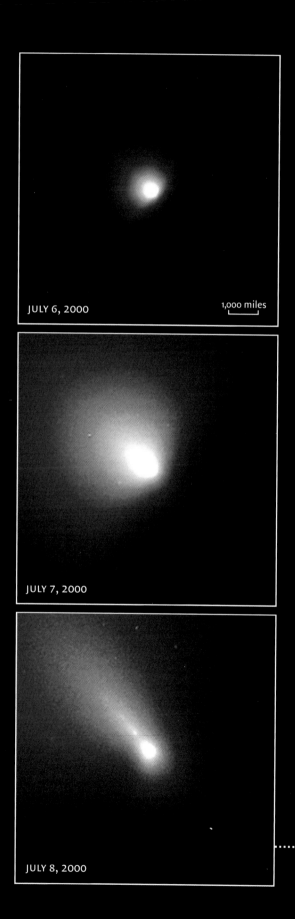

JULY 6, 2000   1,000 miles

JULY 7, 2000

JULY 8, 2000

Piercing the heart of night, a Leonid meteor (above) consumes itself in Earth's atmosphere. The Leonids— named, like all meteors, for the constellations where they seem to originate—shower Earth each year in mid-November. Barely larger than grains of sand, they produce dazzling light shows due to their great speed—around 44 miles a second.

Usually lackluster comet LINEAR achieves brief fame in July 2000, when it blows off a piece of its crust. The Hubble Space Telescope happened to be watching and captured this sequence (left). "We lucked out completely," said one scientist. A mist of ejected dust reflected sunlight, increasing the comet's brightness for several hours. Scientists have several theories explaining the explosion but haven't agreed on one.

Bright underlying ice, exposed for less than 10 million years, gleams smoothly in this artist's impression of Asbolus (below), a 48-mile-wide chunk of ice and dust known as a centaur. Centaurs orbit the Sun between Jupiter and Neptune. Although the Hubble Space Telescope didn't actually see the crater—Asbolus is too small and far away for that—spectral analysis of its surface composition shows a complex chemistry that scientists believe was caused when the ice was exposed by an impact with some other body.

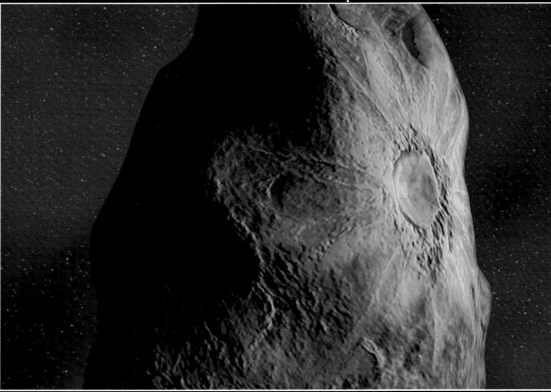

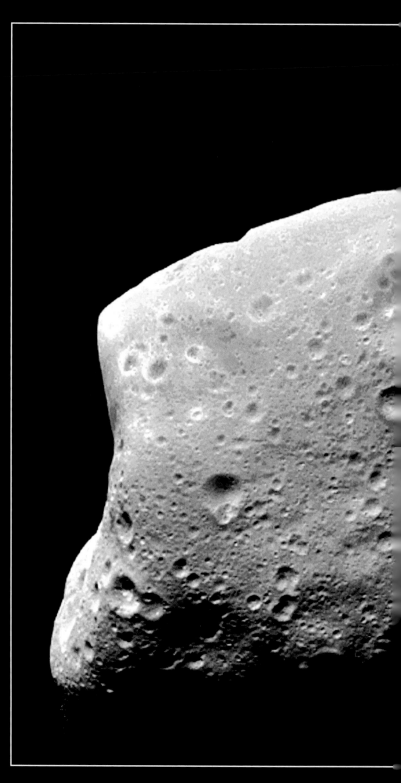

Rare "moonlet"—itself a tiny asteroid—orbits the asteroid Pulcova (above). Only three such asteroidal moons have been discovered. Astronomers atop Mauna Kea in Hawaii captured this image using specialized equipment that compensates for the blurring normally caused by Earth's atmosphere. While Pulcova's diameter is roughly 90 miles, its moon is about a tenth of that; it orbits Pulcova every four days at a distance of about 500 miles.

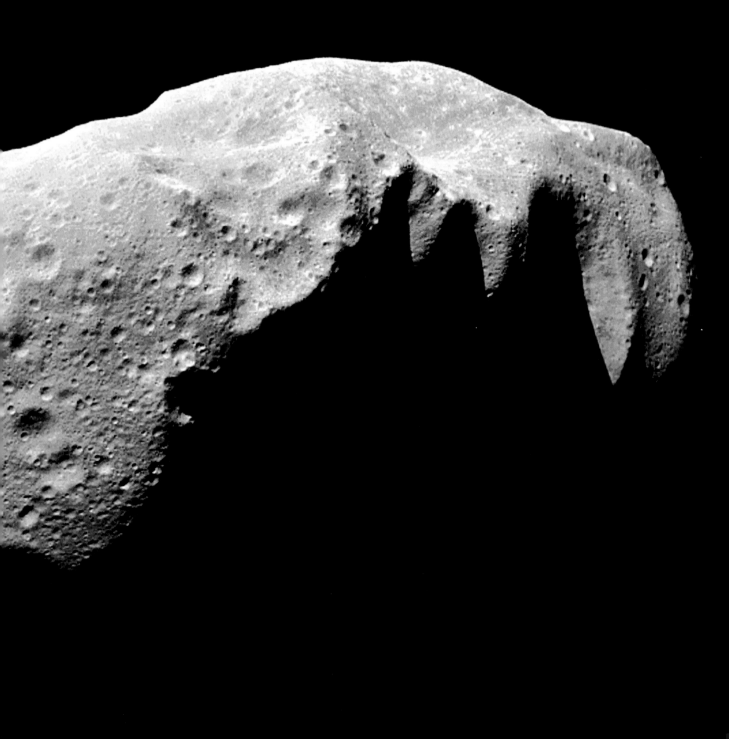

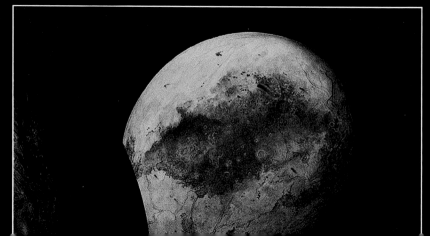

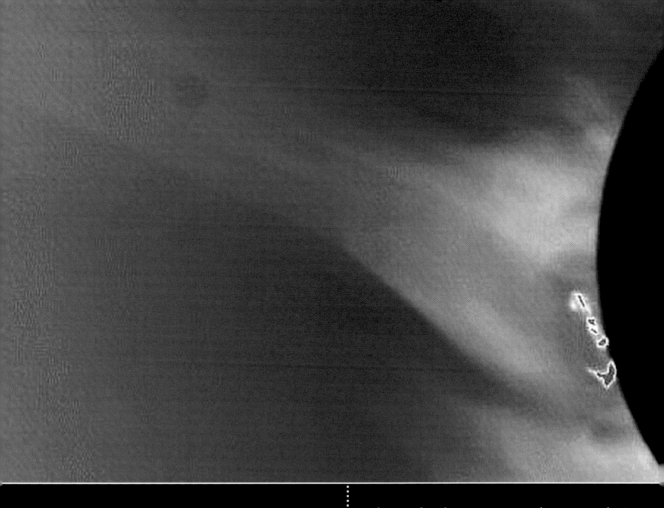

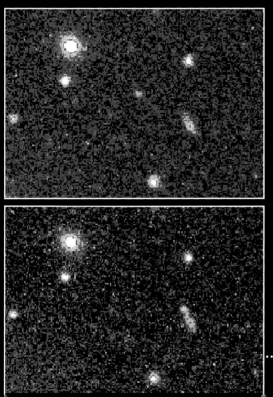

Fiery engine that powers our solar system, the Sun anchors the planets and makes life on Earth possible. This ordinary, average-size star in the middle of its life gives off a range of electromagnetic radiation, some of it deadly. Caplike coronal structures called helmet streamers (above) erupt from the solar surface, their pointed peaks formed by the solar wind.

Beyond Neptune lies the Kuiper belt, a flat doughnut-like ring of thinly scattered lumps of ice mixed with dust and other matter. Scientists estimate that at least 35,000 Kuiper belt objects exceed 100 km (62 miles) in diameter. Pluto and its moon Charon may be two of these objects. A pair of Kuiper belt images (left) taken by the Isaac Newton Telescope, in the Canary Islands, records subtle movements over time.

ONE STAR
AMONG MANY

O ur Sun is fairly typical of the stars that populate the outer disk of the Milky Way. It is a middle-aged star of average mass, luminosity, and diameter holding sway over eight planets—nine if you count Pluto—and a swarm of moons, rocks, and comets embedded in a bubble of space a few light years in diameter.

To the inhabitants of planet Earth, of course, our Sun is anything but average. It is the engine of life as we know it, providing the raw energy that drives our planet's biosphere, shapes its daily weather, and controls its long-term climate. It also is the only star near enough for us to study in great detail; unlocking its secrets provides us with clues and insights about stellar structure and stellar evolution that can be applied to stars across the Milky Way galaxy and beyond.

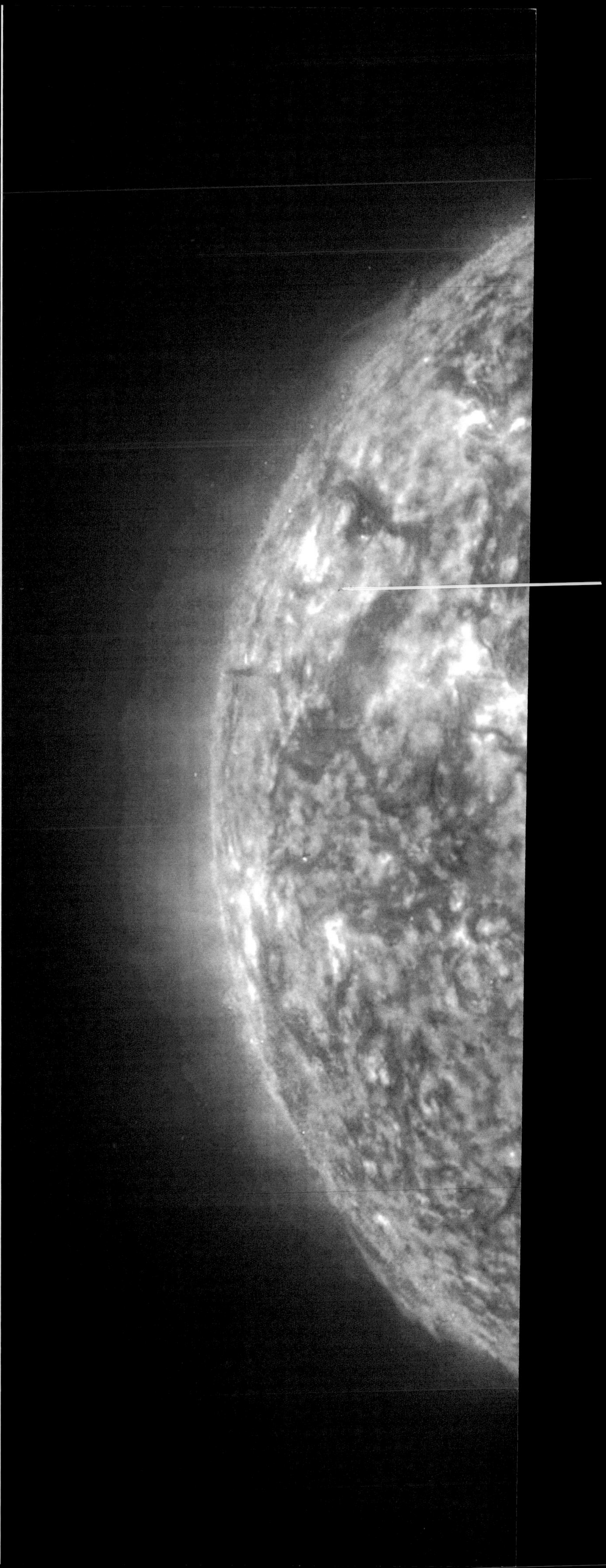

Like a molten handle, a solar prominence erupts. Prominences are clouds of relatively cool, dense plasma suspended in the corona, a layer of the solar atmosphere. The hottest areas of the Sun's face—up to 60,000 kelvins—appear here almost white; red areas are cooler. Solar energy can take millennia to travel from the Sun's core to its surface, but the light it produces takes only 8.5 minutes to reach Earth.

The Sun also is enormous by human standards: 333,000 times more massive than Earth and 109 times as wide, with a diameter of 864,000 miles. Its gravity is so strong that a rocket launched from its surface would need a velocity of 1.3 million miles per hour to escape its clutches. The same rocket would only have to travel 25,000 mph to break free of Earth's gravitational pull.

Even for an average yellow star, the Sun's energy output defies human comprehension. At Earth's distance of 93 million miles (one astronomical unit), a square just one meter (1.1 yards) on a side receives 1,368 watts of solar power. The Sun's total luminosity—the energy received every second across the inner surface of an imaginary sphere with a radius of 93 million miles—is 400 septillion watts. That's 400 followed by 24 zeroes, a number so large it has no meaning in our everyday world.

The source of this enormous energy long puzzled astronomers. In 1871, Hermann von Helmholtz showed the Sun's energy output was equivalent to burning 1,500 pounds of coal per hour over every square foot of our home star's surface. And the Sun's total surface area runs to 2.35 trillion square miles. Helmholtz realized that ordinary chemical energy could not explain this prodigious output. He and William Thompson (Lord Kelvin) speculated that gravitational contraction might do the trick. A star with the Sun's mass could produce its observed thermal energy for 20 million years or so if it experienced gravitational contraction of just 130 feet per year. That seemed reasonable in the mid-19th century, but geologists later proved that Earth—and by extension the rest of the solar system—is far older than 20 million years. Gravitational energy, then, was not the answer.

The solution was nuclear energy, revealed by Albert Einstein's relativity theory and the equation $E=mc^2$, which relates energy to mass. C is the velocity of light, 186,000 miles per second, and because c^2 is such a huge number—34.6 billion—a tiny amount of matter can translate into a tremendous amount of nuclear energy. How much? The current energy needs of the United States for an entire year could be supplied by

Moon's shadow creeps across the Sun during a 1994 eclipse. Since by coincidence the two celestial bodies appear almost exactly the same size in our sky, the Moon completely covers the Sun when its orbit is right. Such solar eclipses occur only about twice a year and never last more than about seven minutes.

converting a little more than one ton of matter. But how is that nuclear energy generated in the Sun?

The Sun, like any other star of similar mass, luminosity, and size, is a spherical ball of hot gas with an internal structure defined by two opposing forces: the inward pull of gravity, which compresses the gas toward the center and causes it to heat up, and the outward pressure that results from the nuclear energy released through that central heating. The delicate balance that exists between these opposing forces is known as hydrostatic equilibrium.

Scientists divide the Sun into three basic layers, plus its atmosphere. The core has a diameter of about 216,000 miles, roughly one-quarter of the Sun's overall diameter. Then comes the radiative zone, which extends about 71 percent of the way to the surface. Beyond that lies the convective zone.

Although the Sun's core accounts for only a small part of the Sun's total volume, it makes up about half of the star's mass. It consists of hydrogen (about 70 percent), helium (about 28 percent), and a smattering of heavier elements that were cooked up by previous generations of stars. But this is not ordinary hydrogen and helium; under the extreme pressures and temperatures that exist in the core, atoms lose their electrons. Free hydrogen nuclei—that is, positively charged protons—whiz about at extreme velocities, crashing into one another in a sort of subatomic demolition derby. So do helium nuclei and free electrons, all of which are so compressed by the weight of overlying layers that they form an ultra-thick soup of electrically charged plasma that is 13 times denser than lead. The core's temperature is some 15.6 million kelvins (28 million degrees Fahrenheit); its pressure is more than 200 billion times Earth's atmospheric pressure at sea level.

Because like charges repel each other, two protons normally would never combine. But at temperatures above 18 million degrees Fahrenheit, they are moving faster than 300 miles per second, packing enough of a wallop to overcome their normal repulsion. Once this barrier is penetrated, the short-range, aptly named "strong" nuclear force takes over, fusing the protons together and releasing energy in the process.

In the Sun's core, three different proton-proton chains of fusion reactions are at work. One accounts for more than 90 percent of the fusion that occurs. In the first step of this process, two protons smash together to form a deuterium nucleus (a proton and an uncharged neutron), a positron (a positively-charged antimatter electron), and a neutrino (a ghostly particle with no charge and little or no mass). In the second step, the deuterium nucleus collides with another proton, yielding a high-energy gamma-ray photon and an isotope of helium that contains two protons and one neutron. In the final step of the chain, two such helium nuclei fuse to produce two free protons and a normal helium nucleus (two protons and two neutrons). Obviously, the first two steps must occur twice to produce the raw material needed for the final reaction.

It turns out that the combined mass of the end products of this three-step reaction is just 0.7 percent less than the mass of the four protons that are needed to create it. It is this tiny difference that is liberated as energy, according to Einstein's famous equation.

By human standards, a second is a mere heartbeat. Yet in a single heartbeat, the Sun turns 700 million tons of hydrogen into helium. In the process, about five million tons of matter—0.7 percent—is converted into pure energy, which supports the star against its own gravitational collapse and, incidentally, nourishes the biosphere of planet Earth. The Sun has been doing this, losing five million tons every second, for four and a half billion years.

"Although the Sun is consuming itself at a prodigious rate, the loss of material is insignificant in comparison with its total mass," writes astronomer Kenneth Lang in *The New Solar System*. "In 4.5 billion years the Sun has consumed only a few hundredths of one percent of its original mass. A more significant concern is the depletion of the Sun's hydrogen fuel" as its nuclear furnace converts the hydrogen to helium. Even so, the Sun so far has used less than half of the hydrogen originally in its core. It will take another seven billion years to burn up the rest.

When that day comes, the Sun will lose its hydrostatic equilibrium. Its core will collapse and its outer layers will balloon outward as it becomes a red giant, incinerating Mercury, Venus, and Earth in the process. Our once brilliant star will become a fading ember, a planet-size white dwarf at the heart of a dead solar system. But all this lies billions of years in the future. Assuming the human family survives that long, it should have plenty of time to find a new home—or to evolve past the point of needing one!

The Sun's middle layer—the radiative zone—is a relatively calm region through which energy released by the core moves by radiation alone. The heat you feel in front of a fireplace reaches you in just that fashion, as photons of infrared radiation. In the convective zone, however, churning clouds of hot gas rise and fall in vast, roiling currents, carrying energy to the Sun's surface by much more efficient convection.

"Although light is the fastest thing around, it does not move quickly from the Sun's core to its visible surface," Lang writes. "Instead it diffuses slowly outward in a haphazard, zigzag pattern, becoming absorbed,

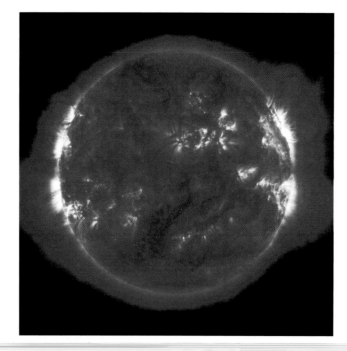

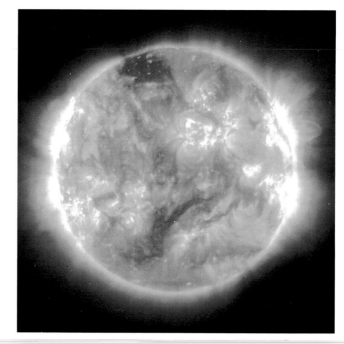

re-radiated, and deflected repeatedly. Because of this continued ricocheting in the radiative zone, it takes about 170,000 years, on average, for radiation to work its way out from the Sun's core to the bottom of the convective zone."

The core and radiative zone appear to rotate as a solid body. The convective zone and the Sun's tenuous atmosphere, however, exhibit differential rotation; that is, points along the equator move faster than points at higher latitudes. Shear forces at the 30,000-mile-thick interface between the steadily rotating radiative sphere and the differentially rotating convective zone appear to power the Sun's titanic magnetic field.

"The Sun's magnetic field arises from the interplay of rotation and boiling convective motions in the outermost 29 percent of the Sun's radius," write Carolus Schrijver and Alan Title in *Sky & Telescope* magazine. "The motions generate electric currents, which create magnetic fields, which in turn help create still more current—the dynamo effect. Once a field becomes strong enough, the pocket of gas holding it rises buoyantly to the surface. The field breaks through with its northern and southern polarities neatly separated into adjacent patches. A large pair of patches is called an active region and is often marked by sunspots."

Sunspots appear as dark splotches against the brighter, slightly hotter photosphere, the visible surface of the Sun. They are caused basically because the Sun rotates faster at its equator than near the poles. This differential rotation causes a shearing and twisting of

the magnetic field that affects the way the ionized solar gas moves. Convection also plays a role in distorting the magnetic field. Large-scale eruptions of gas known as solar prominences typically occur in active regions associated with sunspots. So do solar flares, explosive outbursts of energy that heat localized regions of the solar atmosphere to enormous temperatures and hurl huge clouds of electrically charged atomic particles into space.

A large flare can release the energy equivalent of a two-billion-megaton bomb, write Michael Zeilik and John Gaustad in *Astronomy: The Cosmic Perspective*. Most of that energy escapes into space as ultraviolet radiation or x-rays, which travel at the speed of light and reach Earth in eight and a half minutes. Depending on the magnitude of the flare, such radiation can cause ionization in Earth's upper atmosphere, disrupting communications. Electrically charged particles from a major flare can affect spacecraft electronics and expose astronauts to harmful levels of radiation.

The most powerful transient events in the Sun's outer atmosphere are coronal mass ejections, or CMEs, which blast billions of tons of matter into space at hundreds of miles per second. A CME in the plane of Earth's orbit can trigger a geomagnetic storm, severely buffeting and compressing Earth's magnetic field. This can trigger spectacular auroral displays by causing charged particles to spiral along Earth's magnetic field lines and crash into the atmosphere above Earth's poles. CMEs can induce electric currents in power lines

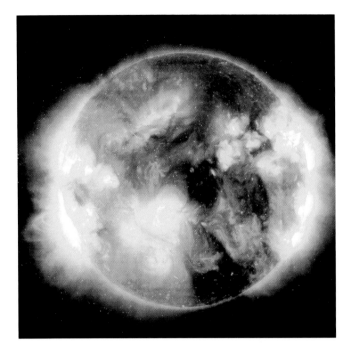

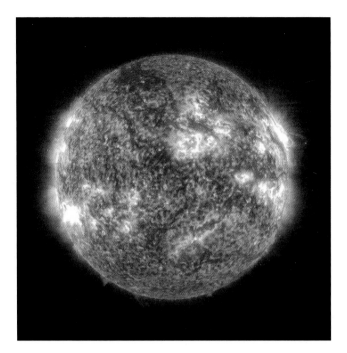

Solar rainbow—four views of the Sun—was created by recording different wavelengths of ultraviolet light, then displaying that data in visible colors. Convection patterns swirl as the blazing corona and interior interact with the relatively cool solar surface. These images come from SOHO, the Solar and Heliospheric Observatory, which operates about a million miles from Earth. SOHO studies the Sun with a dozen different instruments and has even measured subtle seismic waves—Sunquakes—on the solar surface.

on Earth, causing blackouts and other damage. They may or may not be followed by flares.

The number of sunspots rises and falls through an 11-year cycle, with spots first appearing at high latitudes and migrating toward the Sun's equator. The magnetic polarity of spots in the northern and southern hemispheres reverses direction every cycle, so a full cycle actually takes 22 years to complete. But it is neither constant nor automatic; from around 1645 to 1715, virtually no sunspots were observed. During this time, known as the Little Ice Age, Europeans experienced particularly severe winters and cool summers. We now know that was because the Sun is slightly dimmer when sunspots are absent.

Zeilik and Gaustad describe sunspots as "floating islands of electromagnetic storms" embedded in the photosphere, a 300-mile-thick layer of 9,900-degree gas that marks the bottom of the Sun's atmosphere.

Just above the photosphere is the roughly 600-mile-thick chromosphere, visible as a pinkish band during solar eclipses. Solar prominences above active regions originate in the chromosphere and extend high into the self-luminous corona, the outermost layer of the Sun's atmosphere.

Eclipse-watchers know the corona as a billowing crown of light around the Sun during solar eclipses. In a thin transition zone between the chromosphere and the corona, average temperatures jump from 17,500 to more than two million degrees! The corona is hottest wherever interacting magnetic fields are strongest, causing some scientists to believe coronal heating may involve conversion of magnetic energy into heat.

The corona also is the source of the solar wind, a variable stream of charged particles constantly blowing into space. On Earth, it buffets our magnetic field and contributes to auroral displays. In the outer reaches of the solar system, it helps define our Sun's sphere of influence; how far it extends into interstellar space remains an open question. NASA's outbound Voyager probes, more than six billion miles from the Sun as of mid-2001 and still plugging along at more than three astronomical units per year, should reach a point where the solar wind drops below the speed of sound in just a few years. Whether either probe will survive long enough to beam back data about their eventual transition from our solar system to the unimaginable expanse of interstellar space is unknown. But scientists are keeping their fingers crossed.

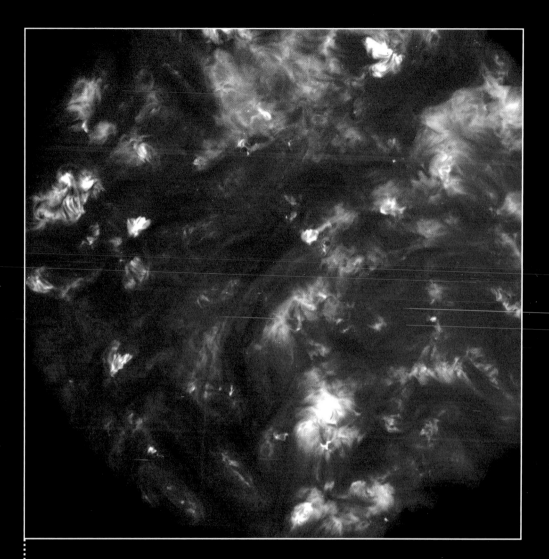

A million times thinner than Earth's atmosphere, the Sun's corona is a boiling maelstrom. This 263-second time exposure captured by TRACE, the Transition Region and Coronal Explorer, reveals the corona's convoluted structure, with loops and swirls reacting to magnetic fields and fierce temperatures.

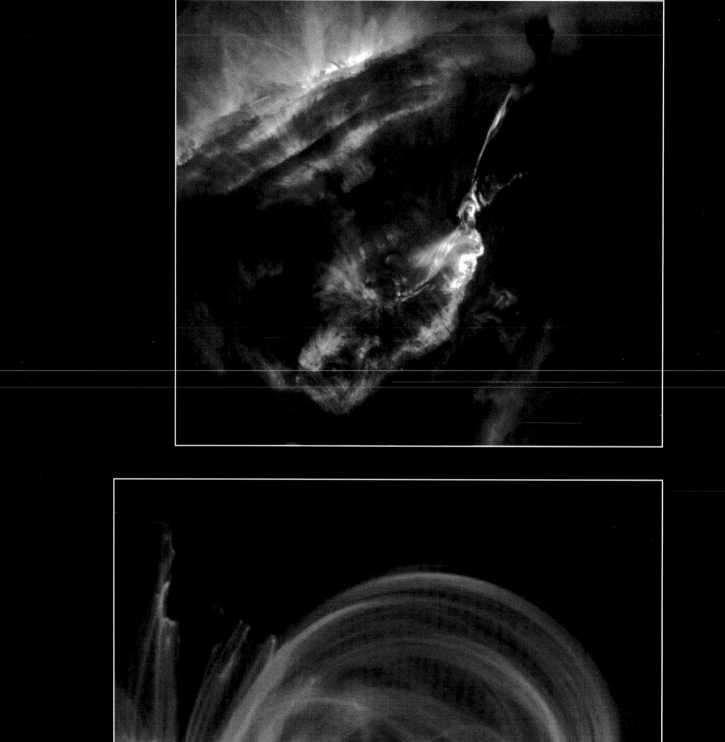

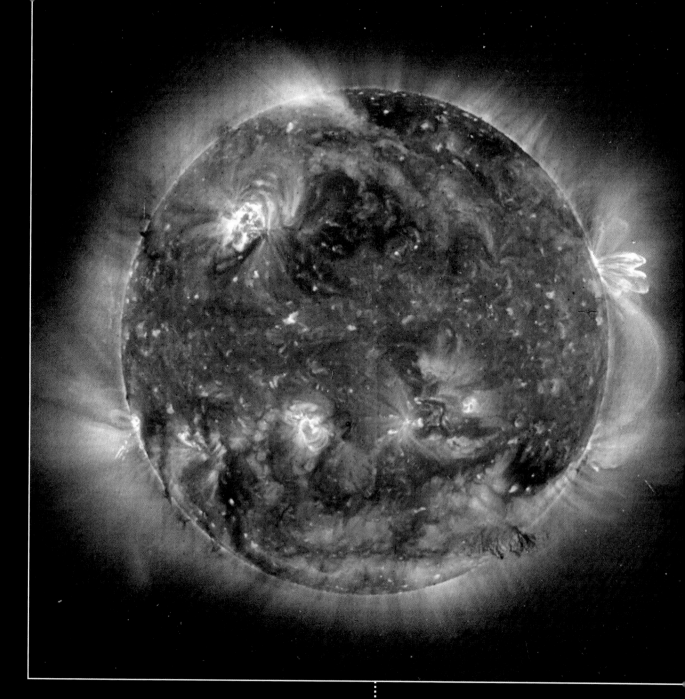

Sun's restless corona ejects a long, wormlike filament of material (opposite, top). Its dark end is relatively cool, the white end much hotter. Recent data from TRACE shows that coronal loops (opposite, bottom) are hottest at their bases, where they emerge from the solar surface. Scientists have long wondered why the corona is roughly 300 times hotter than the Sun's surface. Most heating occurs low in the corona, within about 10,000 miles of the Sun itself.

Arching hundreds of thousands of miles into space, coronal loops (above) can span 30 Earth diameters. This specially constructed image employs three separate wavelengths—each revealing different sola features—to paint a complex but seemingly cool Sun

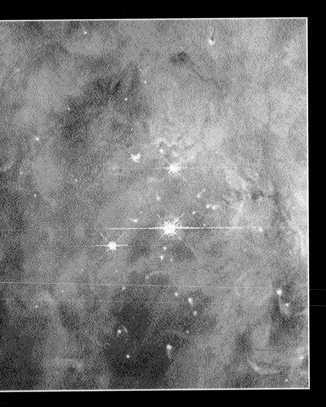

Stellar nursery about 1,500 light-years from Earth,
the Orion nebula (above) shows pale blue and green
in this false-color mosaic combined from multiple
Hubble images. Four massive stars and many smaller,
Sun-like ones appear here. Ultraviolet light from very
young stars—some barely 300,000 years old—bathes
the surface of the nebular cloud.

Jigsaw portrait of a pulsating red giant star (right)
derives from advances in supercomputing and mathe-
matics: After studying convection in the outer layers
of stars for more than a decade, a team from the
University of Minnesota produced this 3-D simulation.
Red giants eventually expand into supergiants, then
blow off their outer layers—or simply explode.

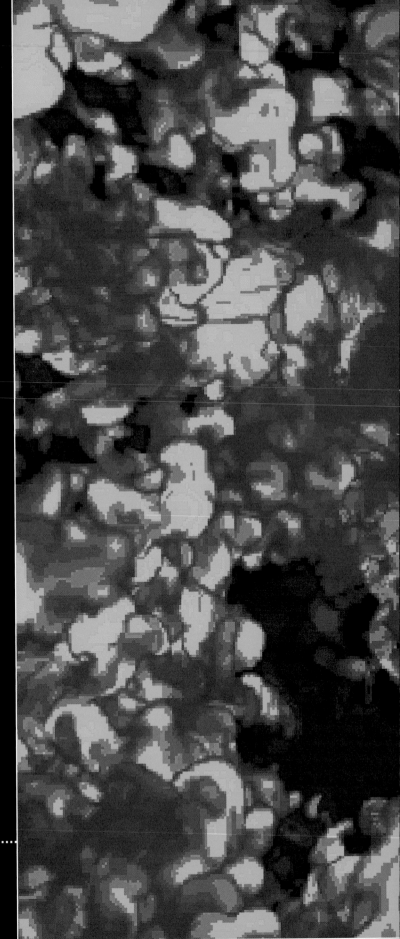

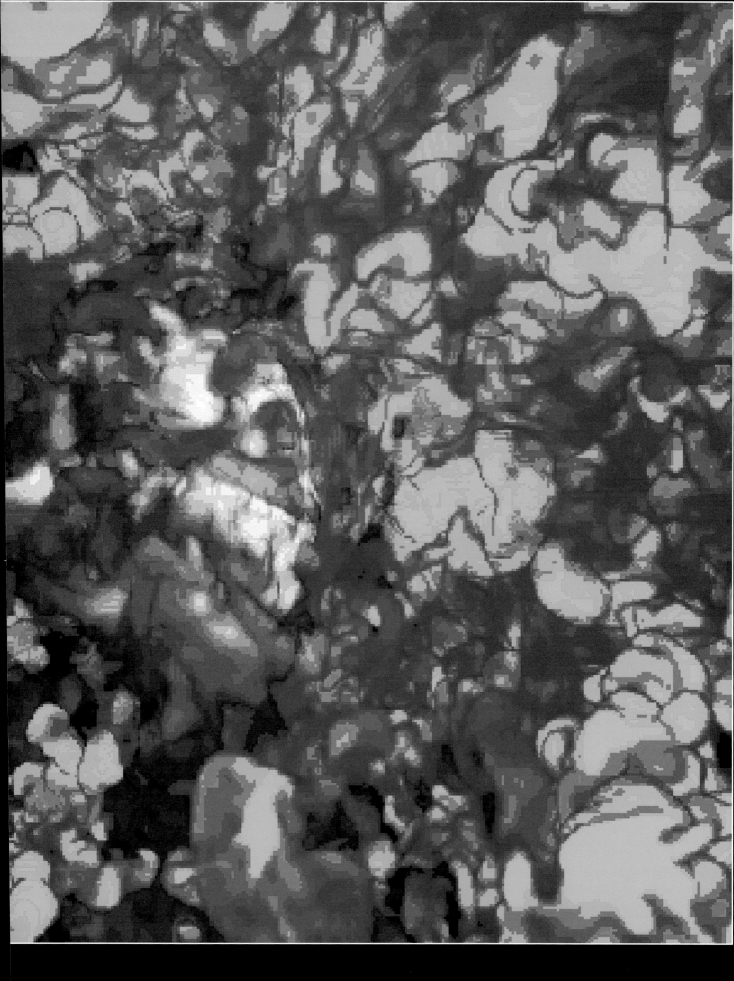

THE MILKY WAY

Most of What You
See at Night

The faint, ethereal band of the Milky Way is difficult if not impossible to make out amid the overwhelming glare of city lights or even the more moderate light pollution that increasingly affects smaller communities. But from a country road or field it remains a mind-expanding sight, the combined glow of countless stars crowding our galactic disk in a glorious tapestry of light, stars that—from Earth's remote vantage point—are too dim, too far away, too obscured by intervening clouds of gas and dust to show up as single points of light. The entire night sky offers only about 2,500 stars that can be distinguished with the naked eye.

Leaving Earth's solar system behind, we venture out into the galaxy proper, far beyond the seemingly vast bubble of space defined by Pluto's orbit, the Kuiper belt, and even the

Deep in the Milky Way, a complex of stars and nebulae called RCW 108 bears fruit: New stars are being born here. Some 600 individual exposures from the European Southern Observatory's La Silla facility in Chile were used to create this, one of the widest, deepest, and most spectacular infrared images ever obtained of a star-forming region. Our disk-shaped home galaxy swarms with perhaps 200 billion stars.

Oort cloud. A viewer 50 light-years from Earth—just around the corner at these scales—would see our Sun as a faint, barely visible point of light amid swarms of equally dim stars spread across the heavens. A viewer even more distant would see these stars as part of an immense, slowly rotating disk some 100,000 light-years across and more than 2,000 light-years thick.

The central region of this disk is the galactic core, some 30,000 light-years across, bulging with old, tightly packed stars around a supermassive black hole masked by thick veils of interstellar gas and dust. It is surrounded by a spherical halo populated by 150 or so known globular star clusters made up of millions of suns 11 to 14 billion years old. The halo has a radius of at least 50,000 light-years. Extending away from the core in symmetric, tightly-bound arcs are the galaxy's spiral arms, seemingly permanent but actually transient manifestations of moving pressure waves that result in countless thousands of star clusters, compressed clouds of molecular debris, and the stellar nurseries that give birth to stars like our Sun.

As if the Milky Way is not difficult enough to fathom, it is part of a still larger structure, a gravitationally bound cluster of more than 30 galaxies known as the local group. The nearest large member of this group, the Andromeda galaxy, is a bigger version of the Milky Way and, at a little over two million light-years away, is the most distant object visible to the naked eye from Earth. And beyond the local group? Clusters and superclusters of galaxies beyond measure extend across space and time as far as the world's most powerful telescopes can see, steadily drifting farther apart as the space separating them continues its relentless expansion in the ongoing aftermath of the Big Bang.

Completely lost to view at this scale, of course, the Sun anonymously orbits the Milky Way's core at a mind-numbing distance of 28,000 light-years out and about 20 light-years above the central plane of the galactic disk. Although it moves 137 miles per second—490,000 mph—relative to the core, the Sun still takes about a quarter of a billion years to complete one circuit of the galactic nucleus.

Two-thirds of the way through its most recent orbit, unnoticed in the daily drama of a galaxy containing 200 billion or more stars and untold billions if not trillions of planets, Earth's dinosaurs were destroyed in a purely local catastrophe that eventually allowed human beings to evolve to the point of sentience and curiosity about their home in space.

But how can beings that evolved on a planet with such finite horizons comprehend the unearthly scale of a structure so colossal it takes light, traveling 186,000 miles every second, 1,000 centuries to travel from one edge to the other? Once again, we must turn to analogies more familiar to human experience. In *Burnham's Celestial Handbook*, a reference for amateur and professional stargazers alike, astronomer Robert Burnham Jr. asks the reader to imagine that the Milky Way covers all of North America. At that scale, individual stars would appear as tiny pinpricks of light separated, on average, by the length of two football fields.

"The solar system, if we can locate the exact spot to look for it, will be about two inches in diameter, and the Sun and Earth will appear as two pinpoint dots about one-thirtieth of an inch apart," Burnham writes. "The Earth, in fact, will be totally invisible to the naked eye on this scale…a sub-microscopic dot a few millionths of an inch in diameter."

It's not a bad idea to read those two paragraphs again. Picture North America in your mind. Then picture a point in, say, Arizona, that is a few millionths of an inch in diameter, a tiny chip off a single grain of sand selected from the continent's vastness. That is Earth against the backdrop of the Milky Way.

Size alone is not the only aspect of the Milky Way that boggles the mind. Consider its mass. Based on the Sun's orbital velocity and distance from the core, astronomers estimate the inner portions of the Milky Way contain more than 100 billion solar masses. But this does not include the mass of stars and dust farther out or the contribution from dark matter, the unseen material believed to make up 90 percent of the mass in the universe. Altogether, the Milky Way may contain well over a trillion times the mass of our Sun.

Viewed from Earth's location in the suburbs of the Milky Way, the galaxy's disk appears as a relatively dim, diffuse band—a milky way—stretched across the night sky. When we look into that band, we are looking into the flattened disk of the galaxy. The galaxy's brilliant heart is located in the constellation Sagittarius, hidden behind intervening clouds of interstellar dust.

Hidden, but not invisible. Radio emissions from the core reveal a group of relatively small objects near the galactic nucleus referred to collectively as Sagittarius A. One of the objects in this group appears to be at or very near the gravitational rotation axis of the galaxy. Powerful x-rays stream from this region; rapidly moving clouds of electrically charged gas appear to be circling a huge concentration of matter. Part of one spiral arm near the core is expanding outward at more than 100,000 mph, propelled by some unknown energy source. That evidence, plus other observations that indicate the source of all this energy must be smaller than a solar system, lead astronomers to suspect a supermassive black hole at the center of the Milky Way, a voracious star-eater perhaps one million times more massive than Earth's Sun.

Observations of Andromeda and other spiral galaxies show that the nucleus and central bulge of each are composed primarily of old, long-lived red stars with relatively small amounts of elements heavier than helium. These are known as population II stars. The 150 or so globular clusters that surround the galactic core in a tenuous halo also consist of population II stars that date back to the birth of the universe, or close to it. The brightest globular cluster in the sky as viewed from Earth is Omega Centauri, easily visible to the unaided eye in the southern constellation Centaurus.

Roughly 17,000 light-years from Earth, Omega Centauri stretches roughly 200 light-years across and includes more than one million stars. In its core, stars are packed two per cubic light-year—as opposed to one star in every 300 cubic light-years, the density in our Sun's vicinity. Through even a small telescope, Omega Centauri makes a magnificent sight, "beyond all comparison the richest and largest object of its kind in the heavens," wrote British astronomer John Herschel in the 1830s. "The stars are literally innumerable." It is visible in the Southern Hemisphere and from the extreme southern United States. But dozens of other spectacular globular clusters are within reach of small telescopes farther north, including M13 in Hercules, second only to Omega Centauri.

Extending from the galactic core of the Milky Way are segments of at least two spiral arms packed with stars, star clusters, softly glowing nebulae, and clouds of dust and gas that include the molecular wreckage of dead suns. In contrast to the core, our galaxy's spiral arms are made up of hot young population I stars relatively rich in elements heavier than hydrogen and helium, elements that were cooked up by preceding generations of stars.

From Earth's position inside the disk of the Milky Way, astronomers can see only segments of these spiral arms. A typical segment might have a width of 1,500 light-years and a thickness of 500 light-years. The Sun is located near the inner edge of a segment known as the Cygnus arm. About 10,000 light-years outward is the Perseus arm, while 6,000 light-years inward toward the core is the Sagittarius arm. There is some evidence of yet another arm closer in, as well as the so-called "expanding arm" circling the central bulge at a distance of about 1,000 light-years, apparently being blown outward at high velocity by energy released in the galactic core.

These arms are populated by stars clumped in clusters large and small or—like our Sun, floating basically alone in a sea of space. Some clusters, like the familiar Pleiades in the northern sky, may seem to contain only a handful of brilliant stars. Actually, they contain hundreds. Others may include thousands. Unlike globular clusters, such open star clusters are generally irregular in shape and are frequently embedded in or accompanied by glowing clouds of gas.

Such nebulae are particularly dramatic. The Triffid nebula in the constellation Sagittarius, for example, glows redly by its own light, as massive young stars heat up clouds of hydrogen gas. At the same time, dust embedded in the cloud reflects the light of stars within it, producing a bluish tint. Thicker bands of dust block out the light entirely. Clouds heated by embedded stars are called emission nebulae, or HII regions, and countless examples abound, including the fabulous Orion Nebula, the middle "star" in Orion's sword and one of the most spectacular objects in the sky. Clouds that merely reflect starlight are known as reflection nebulae. Long-exposure photographs of the Pleiades show hints of such clouds in glowing blue light.

About 10 percent to 15 percent of the visible matter of the galaxy's disk is made up of interstellar gas and dust, the tenuous medium that is the raw material for new generations of stars. The vast majority of this mix is gas, huge clouds of mostly hydrogen, composed of uncharged atoms, ions, electrons, and various molecules. The gas particles generally outnumber their larger dust grain counterparts a trillion to one.

Interstellar dust originates in clouds of gas blown off by exploding stars and in the relatively dense solar winds that spread outward from red giant and supergiant stars. As the gas in the solar wind cools and condenses, tiny grains settle out like frost on a winter morning. The composition of the resulting dust grains depends on the composition of the solar wind and, ultimately, on the nuclear processes at work in that particular star's core.

The spiral arms that define the Milky Way's outer disk are not physical objects. If they were, the arms would have wrapped tightly around the nucleus billions of years ago. Rather, astronomers believe, the arms are the manifestation of density waves that sweep around the core at a constant velocity. Here's a description by Michael Zeilik and John Gaustad from their book *Astronomy: The Cosmic Perspective*:

"This wave produces all the signposts of a spiral arm—young stars, HII regions, dust lanes. None of these objects lasts very long. As they die and the density wave moves, new spiral-arm tracers are born from the interstellar medium. So a spiral arm always contains the same *kinds* of objects but not the *same* objects. Any particular arm is a transient phenomenon. Individual objects rotate at the speed appropriate for their distance from the center, but the wave pattern rotates with a constant angular speed, and does not wind up."

In other words, spiral arms are not permanent structures. The stellar nurseries, the glowing emission nebulae, the hot, young stars that together define what we think of as the Milky Way's spiral arms are strictly transient features, caused by compression as the density wave moves through the interstellar medium.

"The gas in the disk piles up at the back of the wave," Zeilik and Gaustad write. "The buildup of pressure and density heats up the gas suddenly so that a shock wave forms along the front of the density wave.... The compression at the shock squeezes neutral hydrogen clouds together. This shock may initiate the collapse of the clouds to form giant molecular cloud complexes, which in turn create young stars and HII regions. Such squeezing also helps to make dust out of the gas, and a thin dust lane forms along the shock front. The compression of the interstellar medium by the density wave forms the features associated with a spiral arm."

This density-wave model of spiral arm formation explains many features of spiral galaxies, but does not explain how the waves originate or what provides the energy to keep them going for billions of years. A possible explanation is that massive episodes of star formation, gravitational interactions, and other factors combine to create ripples of self-sustaining star formation and transient spiral structures.

The current uncertainty over the structural whys and hows of the Milky Way can be forgiven when one realizes that just three-quarters of a century ago, when my father was a teenager, astronomers were not at all certain that other galaxies existed. In the span of a single lifetime, scientists have measured and mapped the Milky Way with remarkable precision and developed intricate explanations for what makes its billions of stars shine. And my father, perhaps, can be forgiven for smiling when he hears such confident statements. Now in his late eighties, he knows better than most how quickly human pronouncements can change.

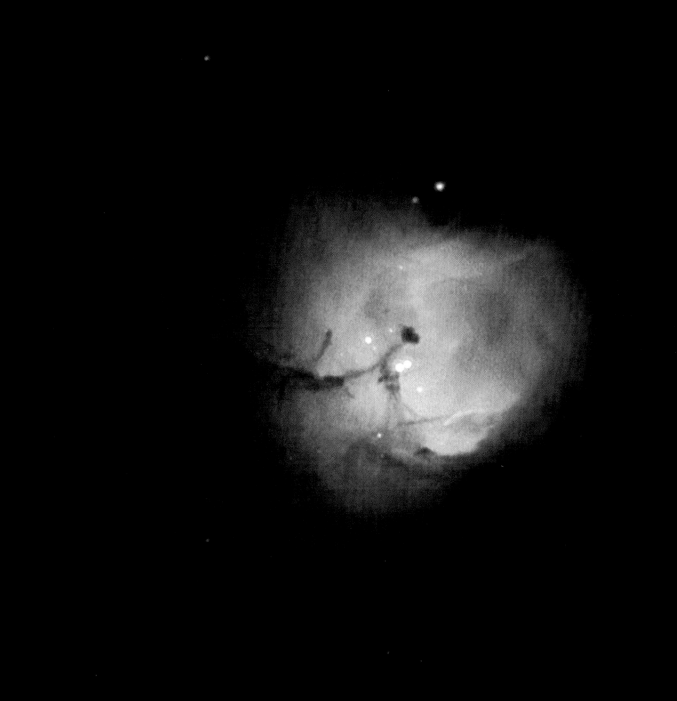

NASA's Hubble Space Telescope finds more new stars forming deep within the Small Magellanic Cloud, a satellite galaxy of the Milky Way and our nearest galactic neighbor. Young, brilliant stars are cradled within a nebula, a glowing cloud of gas. The new stars are losing material at a rapid rate, sending out strong stellar winds and shock waves that are hollowing out a cocoon within the nebula. The brightest stars in this cluster have a luminosity equal to 300,000 of our Suns. Until Hubble, astronomers referred to this nebula as "The Blob," because they could not see such fine detail from ground-based telescopes.

Combining visible and infrared light, Hubble studies distant galaxies (above) in the constellation Tucanae. Reddish galaxies show up in infrared light; bluish galaxies in visible. A brilliant, nearly face-on spiral galaxy appears at upper right. Some of the brightest objects in this view are foreground stars in our own Milky Way galaxy.

Making the first broad search for planets far beyond our local neighborhood, Hubble explored a swarm of 35,000 stars in the globular cluster 47 Tucanae (left) and found none. Some astronomers had expected as many as 17 hot Jupiter-class planets would be discovered here. This failure could indicate that conditions for planet formation and evolution differ fundamentally elsewhere in the galaxy from those we know.

M y best friend is a passionate environmentalist who recy- cles virtually everything and gently urges me to follow

lly separating cans from plastic morning after an evening of star- ccurred to me—as it has to many at my friend's desire to maximize ur limited resources is reflected in ric of the universe.

ht before, I had aimed my 8-inch t the glowing wisps of the Orion marveled at how the galaxy renews ing new stars, planets, and in at se, thinking beings, from stock- ms, molecules, and dust that are being replenished by generations ns. On a galactic scale, as some ic service announcement might say, not just a good idea. It's the law.

r the mythical daughters of Atlas, the glitter brightly in the northern constellation Also called the Seven Sisters, this "open as been recognized since ancient times. clusters of this type, it actually contains of stars, most of which are visible only escope. It lies about 400 light-years from id a wispy cloud of cold dust and gas

Our galaxy's reservoir of raw material can be found in the vast reaches of interstellar space where gas and dust grains—agglomerations of molecules and atoms—float about in immense clouds, or nebulae. The gas is mostly hydrogen and free electrons. While its density is extremely low—better than a hard vacuum by Earthly standards—its volume is so enormous that our galaxy's total amount of gas and dust is equivalent to billions of solar masses.

Glowing by reflected starlight, energized by the searing radiation of embedded stars, or silhouetted against brighter star fields beyond them, the tenuous nebulae sprinkled throughout the arms of the Milky Way come in a rich variety of sizes and shapes, ranging from relatively small, compact clouds to vast stellar nurseries spanning hundreds of light-years.

Nebula—Latin for mist or cloud—is a catchall term initially applied to anything in the night sky that wasn't a comet or a clear point of light, as are the stars and planets. Even ancient stargazers were aware of such hazy, extended objects; the Milky Way itself appears to the unaided eye as a huge cloud of sorts, and a handful of smaller, fainter patches of misty light also were known before the telescope was invented. But as telescopes came into widespread use during the 17th century, stargazers began stumbling across nebulae in bewildering variety. They even discovered that some objects long considered to be stars—including the middle "star" in Orion's sword—were in fact swirling, misty nebulae.

One of the first major sky catalogues to include nebulae was compiled by French comet hunter Charles Messier, in the late 18th century. Still in use today, his catalogue's 105 objects range from galaxies to star clusters and the brighter nebulae, providing a handy

Like the common picnic pest, the Ant nebula creeps across a blanket of nighttime sky. Ground-based telescopes detect only its head and thorax; this Hubble image, ten times more detailed, sees the creature's body as a pair of fiery lobes of gas and dust ejected from a dying central star. Astronomers are struck by the symmetry of this stellar explosion. Perhaps a yet-undetected second star shapes the outflowing gas with its gravitational forces; perhaps stellar magnetic fields play a role. Our home star may create similar dazzling patterns when it finally flickers out.

list of targets for amateurs with small telescopes. Many astronomers of Messier's day believed nebulae were simply groups of unresolved stars, beyond the power of their best telescopes to separate. But in 1864, English astronomer William Huggins used a light-splitting spectroscope to analyze the light from a nebula known as NGC 6543 and was surprised to find that it actually was a cloud of gas.

What were nebulae, and why did one look so different from another? Were different types related? Some appeared to be beautifully symmetrical spiral pinwheels. Others, like the Orion nebula discovered in Orion's sword, seemed to be extended clouds encompassing huge regions of space. Still others, relatively compact and roughly circular, were dubbed planetary nebulae. As it turns out, they're all different things, caused by very different phenomena. But their true nature was not fully understood until the 20th century.

Spiral nebulae were shown to be distant galaxies far beyond the Milky Way. Planetary nebulae proved to have nothing to do with planets; they and supernovae remnants like the Crab nebula are the expanding shells of gas blown off by dying stars. The immense clouds in Orion are actually stellar nurseries at the opposite end of the stellar evolution spectrum. Gigantic clouds of cosmic gas and dust, in other words, are evident both in star birth and star death.

Today, astronomers recognize four general types of interstellar clouds: Cold, dark "molecular clouds" consist of volumes of molecular hydrogen and more complex molecules far below 0°F; warmer, so-called H I regions contain electrically neutral hydrogen, while hot clouds of ionized hydrogen surrounding hot, massive stars characterize H II regions; finally, very hot clouds of gas and dust, superheated by supernova explosions, make up what is known as the hot interstellar medium.

Molecular clouds often form dark nebulae, which are seen in silhouette against brighter backgrounds. The famous Horsehead nebula near the left-most star in Orion's belt is an example. H II regions like the great Orion nebula often give rise to emission nebulae, in which ultraviolet light from hot stars within the cloud itself triggers a reddish glow. Reflection nebulae occur when the light from nearby stars bounces off dust grains in an intervening or adjacent cloud. The Triffid nebula in Sagittarius is a dramatic example of both reflection and emission phenomena.

Planetary nebulae occur when stars like our Sun blow off their outer atmospheres in the process of becoming white dwarfs. The expanding clouds of gas thrown into space by such dying suns often result in spectacular nebulae; flamboyant names such as the Butterfly, Eskimo, Sunflower, Dumbbell, and Owl reflect their shapes. The Ring nebula, in the constellation Lyra, looks like a tiny bluish doughnut when seen through a small telescope; larger instruments resolve it into a thick smoke ring half a light-year across, expanding around a central star at roughly 12 miles

a second. One might assume the cloud is actually spherical and that it appears doughnutlike because we are peering through more gas toward the edges than we are at the center. But recent Hubble Space Telescope images of many planetary nebulae reveal distinctly bipolar, hourglass shapes, indicating the stars blew their outer layers off in opposite directions, not in an evenly distributed sphere. About 70 percent of the planetary nebulae discovered to date show similar bipolar structure.

More violent supernova explosions blow clouds of gas into space in even more spectacular fashion. A 1987 supernova in the nearby Large Magellanic Cloud blew opposing smoke rings into space at six million miles per hour. Just ten years later, the expanding cloud of debris was one-sixth of a light-year in diameter.

For observers in the northern hemisphere, one of the most amazing nebulae of all is the stellar nursery we call the Orion nebula, where stars are in the process of coalescing before our eyes. Some 1,500 light-years from Earth, it has reached about 10,000 degrees Fahrenheit. It is embedded in an even denser molecular cloud, and it glows due to ultraviolet light from a multiple star system at its heart known as the Trapezium: four bright blue stars visible in small telescopes. In actuality, this group—a type of star cluster—contains more than 700 closely packed young suns, believed to be only about a million years old.

Thus the Orion nebula consists of two great clouds, an H II emission nebula powered by the hot young stars of the Trapezium, and a dark dust-rich H I molecular cloud equivalent in mass to about 1,000 Suns. Concentrations in this H I cloud are believed to be areas of active star formation.

In 1992, astronomer Robert O'Dell discovered dusty disks surrounding evolving stars in the Orion nebula. He assessed them as planetary systems in the process of formation, and christened them "proplyds," for protoplanetary disks. Since then, he and fellow astronomer Zheng Wen have found 55 other proplyds in the Orion nebula. Because such disks are difficult to discern, many more almost certainly are present here.

Studying them gives us direct insight into the birth of our own solar system 4.6 billion years ago.

Most stars are within an order of magnitude of our Sun's mass and are born in groups that spread out fairly quickly. But unlike our Sun, most stars do not exist on their own. By some estimates, two-thirds of all stars have detectable stellar companions orbiting a common center of gravity, and many more probably have companions that are too small to be seen. According to one study, 52 percent of the star systems within 13 light-years of Earth feature two or more gravitationally bound stars.

More than 65,000 visual binaries—star systems whose multiple components can be distinguished either by eye, telescope, or camera—have been studied in detail. Another 1,000 or so pairs of stars that cannot be visually separated have been detected and studied spectroscopically. Such spectroscopic binaries are extremely useful because analysis of their motion allows astronomers to accurately determine their masses and, in so doing, their internal structure. Subtle changes in the light from eclipsing binary star systems—in which one star periodically moves in front of the other as viewed from Earth—also enable astronomers to determine the physical size of one or both and glean information about their atmospheric structure.

Astronomers are not sure how binary and multiple star systems form. It could be that fast-spinning protostars that have not yet reached the extreme densities needed to trigger nuclear fusion simply split apart, producing tightly bound stars. Another possibility is that co-orbiting protostars somehow form in the initial collapse of a contracting molecular cloud. A third scenario is that protostars in crowded young clusters collide or gravitationally interact, resulting in stars that orbit each other at greater distances. One study indicates at least half the protostars in a newly formed cluster gravitationally interact at some point, and at least some of the binary star systems visible today probably formed in this manner.

Single stars like our Sun, binaries, and multiple systems often occur in clusters through a process

Sunflower nebula—also called the Helix—blossoms nearer our Sun than any other nebula. Radial blobs or "cometary knots" inside its red shell give the cluster its nickname. Interior green indicates excited oxygen atoms; red is light from nitrogen and hydrogen.

known as subfragmentation. Such star clusters come in three basic varieties: open clusters, associations, and globular clusters.

Open clusters, like the brilliant Pleiades, are generally irregular groupings that span 10 to 65 light-years across and contain 100 to 1,000 stars. Nearly 1,000 open clusters have been mapped across the band of the Milky Way; anyone sweeping across the galaxy's dim outline with binoculars or a small telescope will spot dozens of such concentrated star clouds. Associations are not as tightly packed and generally contain fewer stars spread over a much larger volume. Rich in massive, fast-burning stars, associations dissipate relatively quickly. An example is the well known Trapezium, at the heart of the Orion nebula.

Unlike open clusters and associations, which result from contracting dust clouds in the galactic disk, globular clusters orbit the galaxy's very core in an enormous spherical halo. They also are much more symmetrical and massive, containing 20,000 to many millions of stars arranged in huge swarms. These ancient, relatively static assemblies include stars that first took shape around the time the galaxy coalesced.

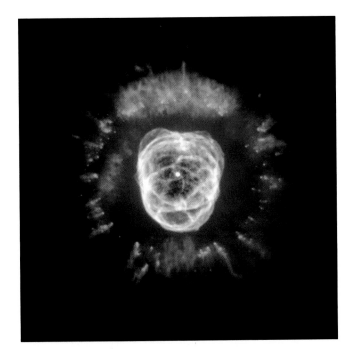

Central star of the Eskimo nebula peeks from a parka-like hood. The first object imaged by the Hubble Space Telescope after a servicing mission in 1999, it boasts a "face" of gaseous material being blown into space during the star's death throes.

"To imagine conditions inside a globular cluster, picture 10,000 stars placed around the Sun at a distance no farther than Alpha Centauri," William Hartmann and Chris Impey write in *Astronomy: The Cosmic Journey*. "If we lived in the core of a globular cluster, our night sky would blaze with starlight ten times brighter than the light of a full Moon!"

In such clusters, the process of star birth is but a distant memory. The real action, galactically speaking, occurs in the galaxy's spiral arms, where cosmic recycling occurs on a truly grand scale as stars are born, grow old, and die, returning at least some of their original material to the interstellar environment.

Most stars, like our Sun, grow old quietly and finally burn out after converting hydrogen to helium for billions of years. This stable hydrogen-burning stage is known as the "main sequence" in stellar evolution. Stars with masses similar to that of our Sun's eventually evolve off the main sequence, becoming red giants and finally white dwarfs, blowing off their outer layers in the process and forming planetary nebulae.

The more massive a main sequence star is, the bigger, brighter, and hotter it will be—and the faster

it will consume its nuclear fuel. A star of one solar mass will remain on the main sequence more than 10 billion years, but a star 10 times as massive will last only 20 million years. Stars even more massive cook up heavier and heavier elements and eventually explode as supernovae. As they blow up, still heavier elements are created and blasted into interstellar space, where they become the ingredients of future stars. In a galaxy the size of the Milky Way, supernova explosions occur every century or so; at this frequency, virtually every point in the galaxy is buffeted by a passing supernova shock wave every few million years.

The debris from a supernova eventually mixes with the gas and dust already present in the interstellar medium, enriching it with heavy elements in the consummate act of recycling. As time passes, supernovae shock waves and other forces increase the density of gas and dust in a given region, eventually triggering gravitational collapse and the birth of a new generation of stars. Thus the ultimate destructive force—a supernova—is also ultimately creative, providing the catalyst for future waves of star birth.

"The most massive stars stay on the main sequence for only the twinkling of a cosmic eye," Hartmann and Impey write. "Some of them evolve into the supergiants region and some less massive ones become ordinary giants. All of them quickly evolve to unstable configurations; many may explode; and all disappear from visual prominence." In so doing, of course, they replenish the interstellar medium and provide the raw material for future generations of suns, planets, moons, and even supernovae. Endlessly.

FOLLOWING PAGES: Named by 19th-century astronomer Sir John Herschel, the Keyhole nebula— also known as Eta Carina—fills this Hubble image, which involved four separate sightings and six different color filters. The Keyhole lies about 8,000 light-years away, has a diameter of about 7 light-years, and contains some of the hottest and most massive stars known.

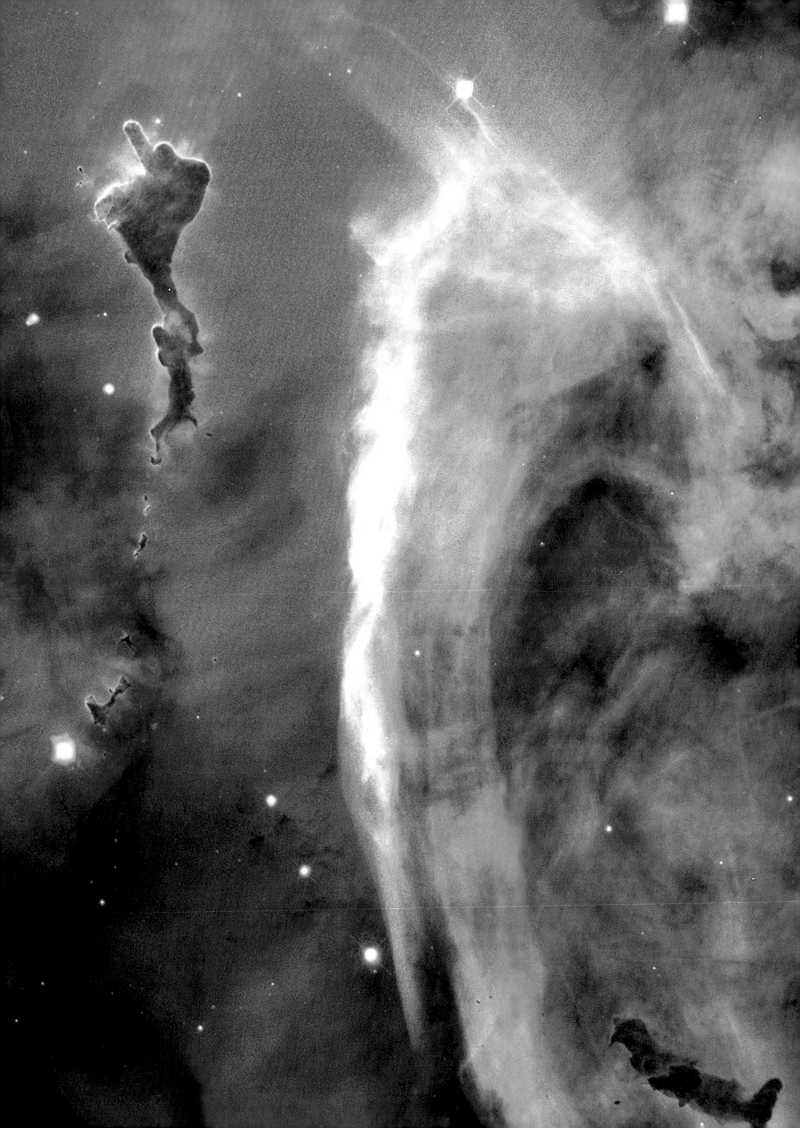

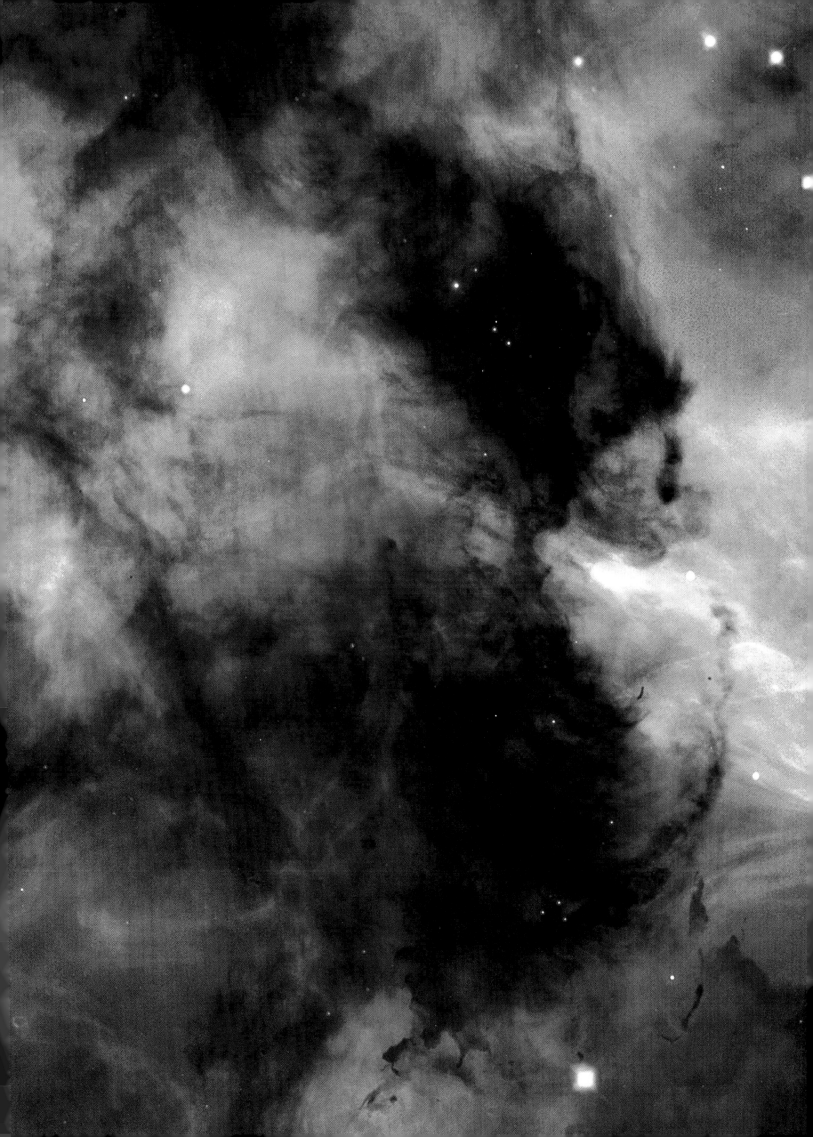

Family portrait in the dusty region between Scorpius
and Ophiuchus (above) is the most comprehensive
collection of nebular types in a single image. Near the
bottom, the dying supergiant Antares steadily sheds
material as it gradually becomes a planetary nebula.
An emission nebula—a gas cloud unassociated with
a particular star but hot enough to give off light—
glows red on the right. Bluish glow near the top is a
reflection nebula, richly endowed with dust and gas.

Brightest of a trio of stars that orbit one another,
the extremely hot HD 148937 (left) continuously
loses mass from its outer layers; especially vigorous
outbursts create the symmetrical shells seen here.

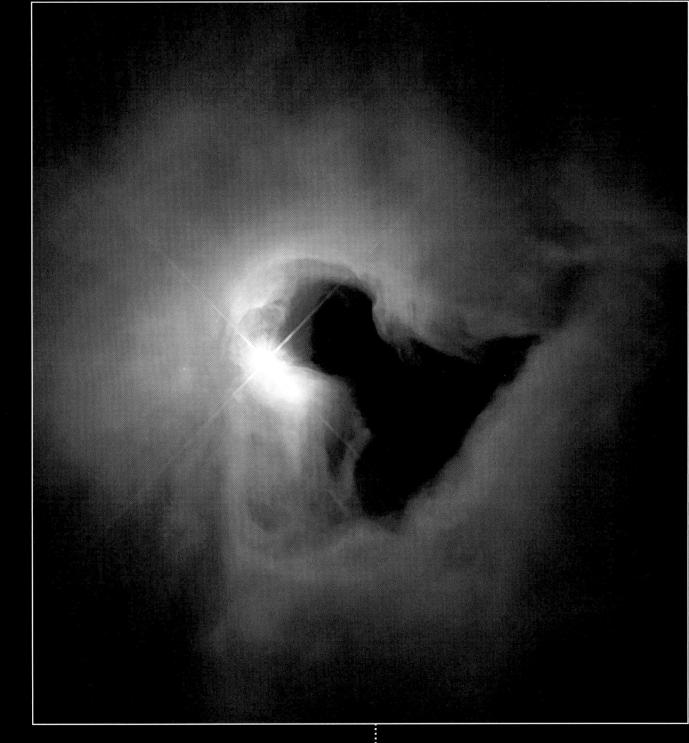

Northernmost "star" in Orion's sword, the nebula NGC 1973-75-77 (opposite) consists not of a single sun but of many. A so-called reflection nebula, it shines as fog does around a street lamp: Not with its own light, but by reflecting the light of other stars. Even so, some component stars are hot enough to excite wisps of hydrogen, which gleam red; starlight scattered by billowing clouds of dust glows blue.

Hubble closeup of NGC 1999 (above), yet another reflection nebula within the constellation Orion, sports a T-shaped jet-black cloud known as a Bok globule. Named for astronomer Bart Bok, such cold clouds of gas and cosmic dust are so dense they block all light behind them. A very hot, bright, recently formed star—V380 Orionis, just to the left of center—illuminates this nebula.

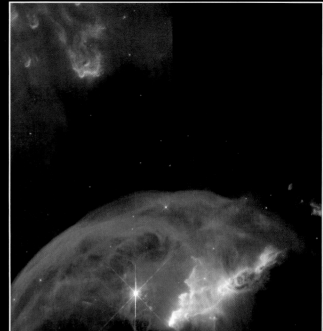

Massive and still unnamed nebula in the constellation Scorpius (left) takes the shape of a swept-back dark cloud, hence its category: dark cloud nebula. The well known Horsehead is another example. The remarkably spherical Bubble (above), a planetary nebula, expands like a party balloon. Its central star, 40 times more massive than our Sun, generates a stellar wind of 4 million miles per hour, propelling particles off the star's surface. Its "bubble" marks the boundary between that wind and its quieter interior; dense fingers of molecular gas near the top have not yet bumped into the expanding shell. The Bubble nebula has a diameter of 6 light-years and lies 7,100 light-years away, in the constellation Cassiopeia.

How Stars
are Born

I n a cartoon by Sidney Harris, two
scientists ponder a blackboard
covered with equations. Linking
one set of equations to another,
bold chalk lettering states: "And then a miracle occurs." One scientist, pointing to the
phrase, tells the other, "I think you should
be more explicit here in step two."

Astronomers dealing with processes that
take billions of years and span distances that
are beyond human comprehension frequently
face similar dilemmas. As Mark Twain once
observed: "There is something fascinating
about science. One gets such wholesale
returns of conjecture out of such a trifling
investment of fact."

How do the huge, brilliant furnaces we call
stars form and evolve? What mechanisms are
at work that create the enormous stellar nurseries we see sprinkled through the galaxy's

Like a troublesome adolescent, a protostar called
Herbig-Haro 32 makes room for itself. Plumes
of gas, appearing here in green, contain material
being shouldered from the protostar's disk into
interstellar space—at speeds of nearly 200 miles
a second. This behavior is typical of youthful stars.
Astronomers George Herbig and Guillermo Haro
did much of the work in this field in the 1950s.

spiral arms? Apart from all its component stars and other bodies, our home galaxy contains huge amounts of interstellar gas—an estimated two billion solar masses worth, but thinly spread over thousands of cubic light-years. Somehow this tenuous gas, so dispersed that it's barely more than a vacuum, is converted into stars, those ultra-dense fusion-powered generators that cook hydrogen into heavier elements of ordinary matter. Interstellar gas. Stars. And in between, a miracle.

In recent decades, using increasingly powerful telescopes and analytical tools, astronomers have made giant strides identifying stars in all stages of evolution, from gravitationally collapsing molecular clouds and dust-swaddled protostars to infant main sequence suns, many with what appear to be retinues of planets in the process of coalescing. Star birth is a complicated process and many unknowns remain. But astronomers are increasingly confident they now understand at least the broad outlines of stellar evolution.

"We need an observational progression of formation from the very beginning right onto the main sequence," writes astronomer James Kaler in his book, *Stars*. "And we have it."

As we've already seen, interstellar space is not a total void. In *The Big Bang*, Joseph Silk explains that this medium "is a complex mix of gas in several phases: ionized, atomic, and molecular. There are giant molecular-cloud complexes and small globules, tenuous very hot gas and wisps of cool atomic gas, and shell-like remnants from ancient explosions of supernovae or winds from massive stars."

Molecular clouds are vast collections of gas and dust that can be 10,000 times as dense as the interstellar average, possessing more than 10 million atomic particles per cubic yard. That may sound like a lot, but compared to Earth's atmosphere, it's basically a hard vacuum. Still, these wispy clouds are dense enough to ensure that their atoms collide more often than not as they are gently buffeted by the density waves that help shape the spiral arms of their galaxy, by gravitational interactions, and by shock waves from supernova explosions. Colliding atoms lead to the formation of

molecular hydrogen and more complex molecules, including carbon monoxide, water, formaldehyde, even the amino acid glycine.

In the core of one huge molecular cloud in Orion, researchers have detected the spectral signatures of more than two dozen different compounds. There is enough ethyl alcohol in a single molecular cloud, for example, to provide cocktails "for a party of enough human beings to populate the entire galaxy," writes Silk. (Galactic bartenders take note: Molecular clouds are so enormous that a spaceship with a funnel more than a half-mile wide traveling at 10 percent the speed of light would need 1,000 years to collect enough alcohol for a single martini.)

How do stars—those dense concentrations of matter—coalesce from such insubstantial material? For stars to form, such clouds must collapse under their own gravity to produce the extreme densities necessary to trigger nuclear fusion. A cloud's initial collapse depends on many factors, including its density, temperature, and the effects of nearby magnetic fields. If a cloud is too warm, the movement of its constituents can provide a sort of pressure that resists gravitational contraction. Although the creation of molecules from atoms provides a cooling effect, ultraviolet radiation from outside a cloud can disrupt this process. If enough dust is present, however, the ultraviolet light will be blocked somewhat, and atoms will more readily link into molecules. Molecules, in turn, radiate thermal energy more effectively than atoms, allowing the gas to cool to the point where gravitational forces can dominate.

In addition to thermal pressure, magnetic effects also tend to oppose contraction. But they can be overcome by outside forces, like the compression caused by a passing supernova shock wave or the sudden removal of a magnetic field.

Given the complexity of interactions that trigger the initial collapse, nearly all such clouds end up with some amount of rotation, typically more than enough to break up a coalescing star. As they contract, they must rotate faster and faster, conserving angular momentum just as spinning ice skaters do when pulling in their arms. Often, a contracting cloud splits apart to form binary or multiple star systems.

For an isolated star to survive, the contracting cloud must shed more than 99 percent of this angular momentum. One way is through magnetic braking, in which energy is transferred to the surrounding environment by magnetic fields. Another way is to transfer the energy to its surrounding disk. Such flattened circumstellar disks, the birthplaces of future planets, appear to be common around young stars.

"In theory, all young stars should have either a circumstellar disk or a companion to take up the cloud angular momentum," write astronomers Robert O'Dell and Steven Beckwith in the May 1997 issue of *Science* magazine. "Observationally, disks and companion stars are common and young binary systems often have disks surrounding one or both stars as well."

As the cloud contracts but before it reaches the density levels required for nuclear fusion to take place, an observable protostar appears. Protostars can be embedded in large nebulae, like the stellar nursery in Orion, or in much smaller molecular clouds. In any case, once a collapsing cloud contracts to roughly stellar dimensions—that is, a few astronomical units across—"the atoms of gas bump into each other frequently enough to produce substantial outward pressure, so the collapse is slowed," write William Hartmann and Chris Impey, in *Astronomy: The Cosmic Journey.* "At this point, we can speak of the object as a pre-main-sequence star."

Stars on the main sequence, of course, develop cores so dense that hydrogen fuses into helium in a stable process spanning millions to billions of years, depending on their size. Pre-main-sequence stars also generate heat and maintain hydrostatic equilibrium, but not initially as a result of nuclear fusion. Rather, stability is first achieved through a phenomenon known as Helmholtz contraction. As molecules fall inward and collide with other particles, temperatures increase to a few thousand degrees, producing a pressure that slows the overall collapse and causes

the formative star to emit visible radiation. Temperatures continue to rise during this slow but steady contraction until fusion reactions eventually begin.

A star's evolution before, during, and after the main sequence depends—like so many things in astronomy—on mass. Low-mass stars can take more than 100 million years to evolve onto the main sequence, while our Sun gets there in a few tens of millions of years. A five-solar-mass star makes the evolutionary jump in just a million years or so. And a star fifteen times more massive than our Sun reaches the main sequence in just a hundred thousand years.

As protostars evolve toward the main sequence, they are typically surrounded by dense clouds of gas and dust that can act like cocoons, blocking visible light. These infant stars are seen primarily through infrared emissions and are thus known as infrared stars. The obscuring dust eventually dissipates, revealing the central star inside.

The link between such infrared stars and the main sequence is a class of objects known as T Tauri stars, named after a variable star that was discovered in the constellation Taurus in 1943. T Tauri stars typically range from 0.2 to 2.0 solar masses and exhibit extensive convection. In many cases, they are associated with huge interstellar clouds like those in Orion. They are young, typically between 20,000 and 1,000,000 years old, and they exhibit rapid changes in brightness. X-ray telescopes detect ten or more major flares a day.

T Tauri stars are surrounded by thick disks of dust that were either ejected from the central star or left over from its initial collapse. This material absorbs short-wavelength radiation from the central star and re-emits it as infrared radiation. Such disks can be several hundred astronomical units in diameter, providing a reservoir of material that ultimately falls onto the star. At the same time, T Tauri stars blow immense amounts of material back into space. Observations show that the equivalent of 0.4 solar mass of material may be blown away as a T Tauri star evolves. Initially interpreted as a strong solar wind, this phenomenon was found to be much more complicated.

Imagine a star surrounded by a rotating circum-stellar disk of turbulent gas. Interactions between the orbiting particles "lead to a slow spiraling of material through the disks onto the star," write O'Dell and Beckwith. "The accretion of matter releases energy. The release of energy heats the star and the disk and is believed to provide the ultimate source of power for the jets that so commonly accompany star formation."

Such jets shoot out from the polar regions of many, if not all, pre-main-sequence stars. The precise mechanism leading to the creation of these bipolar outflows is not yet known, but most theorists believe magnetic fields threaded through the star and its circumstellar disk produce dynamo forces that generate and some-how focus the outflowing material into jets.

"Stars do not just accrete matter, they also spew it out, generating winds that flow away from the star-disk system into the surrounding cloud," O'Dell and Beckwith continue. When the jets plow into surrounding clouds of gas, they create compact nebulae, or areas of hot plasma, known collectively as Herbig-Haro objects, named after two astronomers who found the first example. Several hundred Herbig-Haro objects have been discovered to date and they are now believed to be common manifestations of young stars interacting with their surroundings.

Putting the observational pieces of the puzzle together, astronomers now identify four main stages of star formation. First, slowly rotating density cores form in a molecular cloud. A region of higher density collapses into a gravitationally heated protostar, surrounded by a disk of dust and an even larger sphere of infalling material. A stellar wind develops, along with polar outflow jets. Finally, the dusty remnants of the developing solar nebula thin out, revealing its central star or stars as well as whatever planets may be taking shape from the original disk.

When a star enters the T Tauri stage, its energy is being provided by Helmholtz contraction. As pressure and density increase, the core heats up; eventually fusion reactions begin, counteracting the gravitational contraction. At this point, the star has finally reached the main sequence in its evolution, where it will remain for billions of years.

Michael Zeilik and Stephen Gregory, in *Introductory Astronomy and Astrophysics*, note that bipolar outflows from solar-mass T Tauri stars imply that their disks often extend 100 astronomical units or more in size—comparable to the scale of our solar system—thus they are "just what you would expect for nebulae out of which planetary systems can form."

Stars tend to form in groups, and T Tauri stars are frequently found clumped together in star-forming regions like the Orion nebula. Protoplanetary disks—proplyds—also abound here, revealing infant solar systems taking shape before our eyes. Our own solar system must have gotten its start in a similar fashion.

"Everything indicates that the bodies of the planetary system formed at the same time as the Sun and that they were born from a spinning disk," Kaler observes in *Stars*. "What disk could it be other than the one that we have seen develop all the way from the molecular clouds through to the T Tauri stars?"

If the number of proplyds in the Orion nebula are any indication, planets may be common there indeed. Or maybe not! Recent observations by the Hubble Space Telescope indicate that young stars in the Orion star birth region may not have enough time to evolve planets before their protoplanetary disks are blown apart by intense radiation from hot, giant stars at the heart of the nebula.

Detailed observations show that larger-than-expected dust grains are, in fact, sticking together in these protoplanetary disks. But "these bright stars are trying to tear everything apart," Hubble co-investigator Henry Throop said in a NASA news release. "Which one wins is really a big question. It's like trying to build a skyscraper in the middle of a tornado."

Such massive, fast-burning stars are a natural consequence of cluster formation. If the disruptive effects observed by Throop and co-investigator John Bally are the rule and not the exception throughout the universe, extrasolar planets—and life itself—could be more rare than science fiction fans might wish.

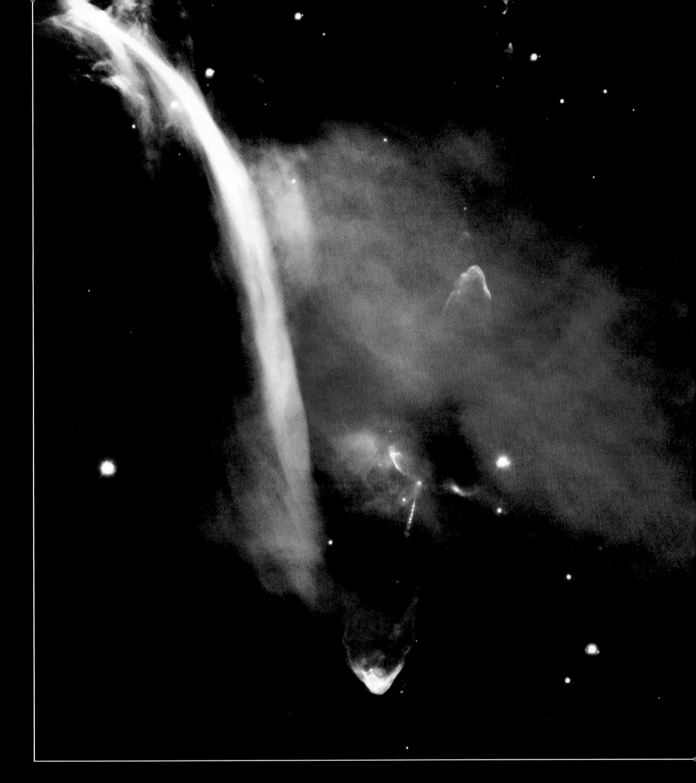

As puzzling as it is spectacular, the large arc of gas known as The Waterfall eludes simple explanation by astronomers. It occurs in a region of Orion near the protostar Herbig-Haro 34, which ejects "bullets" of high-energy particles, one of which shows up here as a bright red streak. Evidently, when gas from H-H 34's surrounding disk collapses onto this young, formative star, a burst of similar particles rebounds like a splash. This particular bullet measures about one light-year in length and is traveling at some 155 miles a second; each end of it wears a glowing cap. The Waterfall lies some 1,500 light-years from Earth.

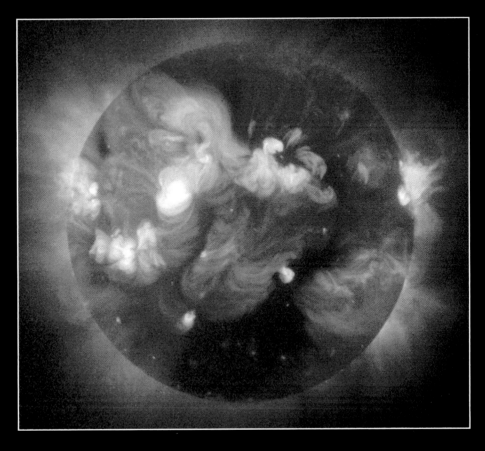

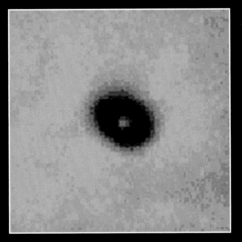

A star is born: In an early stage of stellar evolution, a T Tauri star condenses inside a molecular cloud (opposite). Spinning rapidly in this artist's conception, its dark protoplanetary disk attracts matter as it grows, and its powerful magnetic field generates currents in the coalescing material, melting some of it and spawning monstrous flares. More than 500 T Tauri stars have been observed, some mere babies (left). Young ones remain unstable until their interior temperatures grow high enough to support thermonuclear reactions. Our Sun (above) went through similar evolutionary changes in its youth.

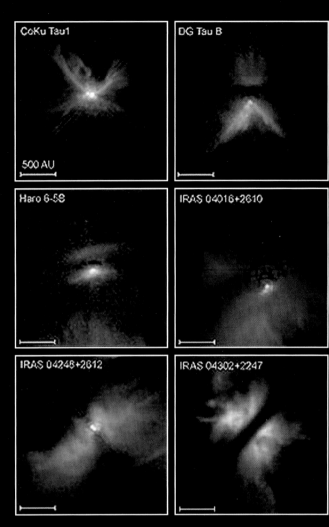

CoKu Tau1	DG Tau B
IRAS 04016+2610	
Haro 6-58	
IRAS 04248+2612	IRAS 04302+2247

500 AU

Star IRS4 spreads its wings like a newborn butterfly, creating a gaseous nebula called Sharpless 106 (left). The nebula spans about two light-years and lies some 23,000 light-years distant, toward the constellation Cygnus. Viewed by Hubble, six extremely young stars 450 light-years away in the constellation Taurus (above) exhibit dramatic disks of encircling dust— an early step in the formation of planets. Dark clumps and bright streamers indicate that some material is still falling into these disks and jets of gas are being ejected, as the stars continue to evolve.

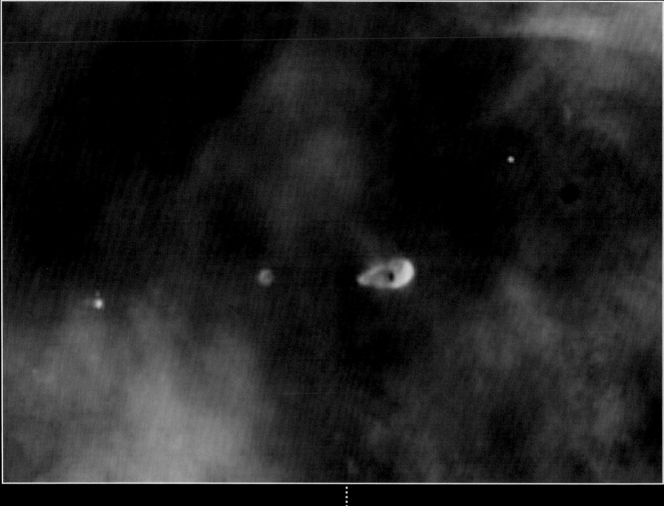

Hubble's sharp eye resolves newly forming stars (above) amid disks of dust and gas in the Orion nebula. Astronomers believe such protoplanetary disks—proplyds—are infant solar systems.

Donut-shaped dust cloud (left) surrounds the star LkHa101. The star itself, too small to be seen, is in the reddish void near the center; its heat has vaporized surrounding dust to create the hole.

Tiny fingerlike projections texture the upper reaches of this column of cool molecular hydrogen gas and dust in the Eagle nebula (opposite). Actually, each of them—dubbed Evaporating Gaseous Globules, or EGGs—exceeds our entire solar system in size. Such stellar "eggs" are bared as ultraviolet light from nearby hot stars slowly erodes the pillar.

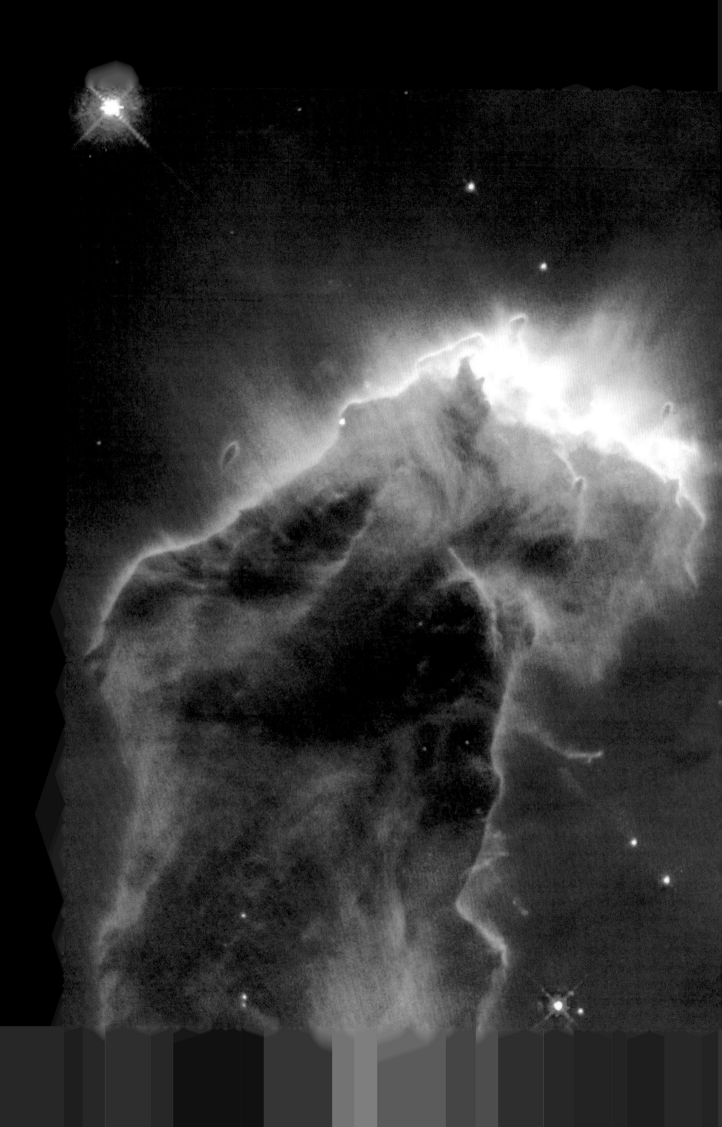

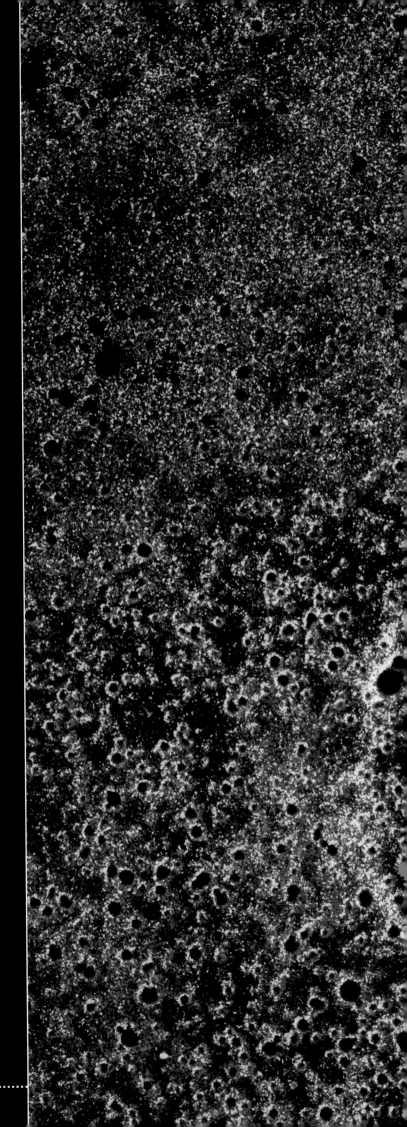

STARS DYING ENDLESSLY

White Dwarfs, Neutron Stars, and Black Holes

T his is the way the world ends. Not with a bang, as a cinder. In the three seconds it takes to read that parody of T.S. Eliot, ur Sun transmutes some 2.1 billion tons of ydrogen into helium. In the process, 14.7 nillion tons of matter simply vanish, directly ecoming the energy that, among other nings, makes life possible on planet Earth. n the 20th century alone, the Sun's core lost 5.5 quadrillion—15,500,000,000,000,000— ons of matter.

The Sun has been consuming itself in this ashion for the past five billion years, burning ydrogen to produce the radiation pressure eeded to counteract the relentless inward pull f its own gravity. Despite this enormous rate f fuel consumption, astronomers believe the un has enough hydrogen left in its core to emain a relatively stable main sequence star or another seven billion years.

Cosmic bulls-eye, Supernova 1987A sprouts rings following the explosion of a massive star in the Large Magellanic Cloud, in February 1987. These rings, echoes of that blast, are the focus of intense study as they continue to evolve. Astronomers created this image by digitally subtracting a photo taken before 1987 from one taken later.

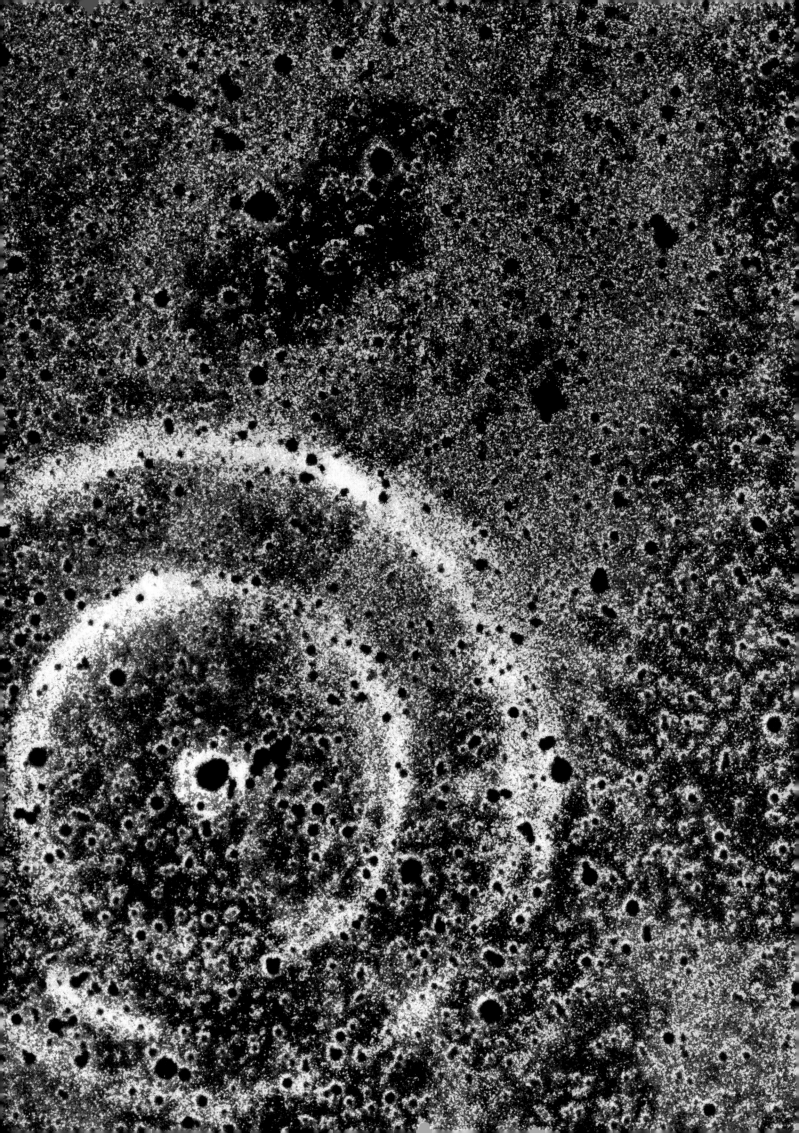

But as the Sun's core steadily loses mass, its temperature and density must increase to maintain hydrostatic equilibrium. When the Sun was an infant, it was slightly smaller, cooler, and dimmer than it is today. As its rate of nuclear fusion slowly increases in the aeons ahead, it will continue to expand and more energy will radiate from it. In other words, our Sun will get brighter. That's not an immediate problem for a star. But it's a death sentence for planet Earth.

Over the next several hundred million years, the Sun's increasing temperature will accelerate the evaporation of Earth's oceans, driving up the opacity of our atmosphere and eventually triggering a runaway greenhouse effect. Ultraviolet radiation will break down the water molecules in our atmosphere and the component hydrogen in H_2O will escape into space. Earth will begin to resemble Venus.

Within a little more than one billion years, Earth's once teeming oceans will disappear as surface temperatures soar above the boiling point of water, turning a once-verdant world into a lifeless, burned-out cinder. Still, our slowly brightening Sun will remain on the main sequence some six billion years beyond Earth's hellish demise, before finally exhausting the hydrogen in its core. At that point, the core's proton-proton chain of fusion reactions will stop, the radiation pressure that supports the star's overlying layers will suddenly drop, and the core will collapse. Hydrogen in a shell around the core will continue burning, but the core's collapse will heat this shell and cause it to expand. As the radius of the Sun grows, its surface temperature will consequently drop off. But the Sun's luminosity, the rate at which energy radiates from a star, will increase dramatically and the Sun will become a bloated red giant.

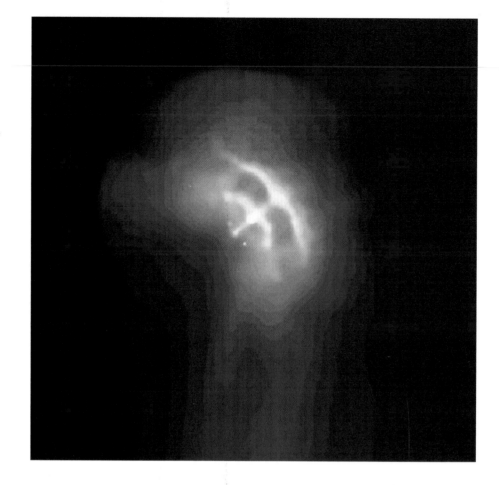

As precisely aimed as the crossbow it resembles, this x-ray image of the Vela pulsar points to the upper right, as does its "arrow," a particle jet emanating from its core star; the arcs are edges of tilted rings, like shock waves. Its heart is a neutron star, the collapsed remnant of a supernova that exploded 10,000 years ago and likely was bright enough to be seen from Earth in daytime. Only 12 miles in diameter but extremely dense, Vela makes 10 rotations a second.

"The effect as viewed from an orbiting planet—could anyone survive to see it—would be awesome," writes James Kaler in his book, *Stars*. "Reddish and bloated, it will appear 50 degrees across in the Earth's sky, quadruple the angular size of Orion and (if we ignore the inevitable slowing of terrestrial rotation) will take over three hours to rise and set." The Sun's core—now only planet-size—has a temperature of nearly 90 million degrees Fahrenheit. It resists further compression not by fusion but by a quantum mechanical effect known as degenerate gas pressure, in which electrons with identical properties cannot be crammed closer together. Eventually, gravitational contraction pushes temperatures high enough to trigger the fusion of helium nuclei into carbon.

Because the core is a degenerate gas, this temperature increase does not immediately cause the core to expand. Instead, it causes the rate at which helium is converted into carbon to accelerate in a burst called a helium flash. Eventually, core temperatures reach more than 600 million degrees, the electrons become "nondegenerate," and the core expands and cools off a bit as helium transforms into carbon. Hydrogen burning continues in a shell around the helium-burning core.

But the helium burns relatively quickly. Once it's gone, fusion stops and the electrons in the Sun's carbon core become degenerate. Once again, the star expands, this time to truly gargantuan proportions. Just how big is open to question, but recent calculations indicate its outer layers may reach or even exceed Earth's present orbit, writes Kaler. "If it overtakes our now-molten planet, the Earth will actually orbit inside the [Sun's] low-density envelope. Friction with the gases will cause it to lose orbital energy and spiral inward, eventually to be utterly destroyed along with Venus and Mercury. Mars likely will be spared, and the heat may be enough to render conditions springlike on the outer planets."

During this second period of expansion, nuclear energy will be supplied by shells of hydrogen and helium, which will shut down and re-ignite in a self-reinforcing feedback loop. As the end approaches, explosive helium flashes will come closer and closer together and the Sun's luminosity will rise and fall up to 50 percent over periods as short as a few decades. These explosions will spawn a superwind that will blow away the Sun's outermost layers, creating a spectacular planetary nebula.

In the Sun's interior, nuclear reactions finally will grind to a halt. The naked core that is left behind will have about six-tenths of the Sun's original mass jammed into a hot, ultra-dense sphere about the size of Earth. This is not massive enough to generate the gravitational energy needed to fuse carbon into heavier elements. And so, less than 100,000 years after our Sun's second period of expansion begins, its core will become a white dwarf, a stellar ember that slowly dims as its pent-up heat radiates away over billions of years.

Such is the fate of the vast majority of stars in the universe, those between 0.1 and 8 times as massive as our Sun. Smaller stars never reach the main sequence at all, becoming degenerate before fusion reactions can begin. These are brown dwarfs. Stars between 0.1 and about 0.4 solar masses fuse hydrogen into helium but never get hot enough to fuse helium into carbon; they end up as helium-rich white dwarfs. Stars like our Sun evolve into carbon-rich white dwarfs, while slightly heavier ones develop the heat to fuse carbon into oxygen; they become oxygen-rich white dwarfs.

The very rare stars with masses between 8 and 30 times that of our Sun face a different fate. Because they burn so hot and fast, they can fuse heavier and heavier elements as they struggle to maintain their hydrostatic equilibrium. They end up with compact cores of solid iron, surrounded by shells of lighter elements. A 15-solar-mass star, for example, can wind up with an iron core surrounded by shells of silicon, oxygen, carbon, helium, and a thick outer layer of hydrogen and helium. Different fusion reactions can proceed simultaneously on the surfaces of different shells.

Fusion of hydrogen into helium goes on for millions of years, but the production of heavier elements like carbon, neon, oxygen, and silicon takes place in stages lasting less than 1,000 years each. The end of the road occurs when temperatures in the core exceed 1.8 billion degrees Fahrenheit, triggering reactions that produce radioactive nickel nuclei. Nickel-56 quickly decays into cobalt-56 and then into iron-56. Because iron nuclei are the most stable in the periodic table, they do not fuse. At this point, fusion stops for good; seconds later, the core collapses at nearly 25 percent of the speed of light, causing the once-huge star to explode in a catastrophic supernova.

This time, there's no white dwarf. Stars with more than eight times our Sun's mass end up with planet-size cores containing at least 1.4 solar masses, a ratio known as the Chandrasekhar limit, named for Indian astronomer Subrahmanyan Chandrasekhar. Such stars have enough concentrated mass to overcome even electron degeneracy pressure, and their cores do not stop collapsing until they reach densities comparable to those of atomic nuclei. When a stellar core reaches this unimaginable state it rebounds, smashing into other material that is still falling inward. The resulting shock wave blows the star apart and, in a process known as explosive nucleosynthesis, creates elements heavier than iron. These elements are blasted into interstellar space where they become the raw material for future generations of stars and solar systems.

The core, meanwhile, has collapsed to the point where protons and electrons merge, leaving a sphere of neutrons supported against further collapse by quantum effects similar to those responsible for white dwarfs. The result is a neutron star, the densest object in the known universe, packing the mass of our Sun into the volume of a small asteroid. A thimbleful of it would weigh some 100 million tons on Earth.

Physicists Fritz Zwicky and Walter Baade predicted the existence of neutron stars in 1934, just two years after neutrons were discovered, writing that "with all reserve we advance the view that a supernova represents the transition of an ordinary star into a neutron star, consisting mainly of neutrons. Such a star may possess a very small radius and an extremely high density." J. Robert Oppenheimer and others developed a theoretical framework to explain neutron degeneracy and establish the mass limits for a neutron star. Ultimately, their work convinced many astronomers such stars must exist. But neutron stars were not detected until 1967, when radio astronomers discovered what later was determined to be a spinning neutron star, or pulsar. Hundreds of pulsars have since been located, all believed to be the remnants of massive stars.

Neutron stars apparently have a thin, solid crust made up of free electrons, neutrons, and a smattering of heavy elements. Just below is a layer of degenerate neutrons that move as a superfluid, that is, without friction. In the core, neutrons and a few free protons move in a frictionless, superconducting state with no electrical resistance. Not surprisingly, anything hitting a neutron star would release enormous amounts of energy. Collisions between neutron stars may, in fact, be responsible for gamma ray bursts, those ephemeral but titanic explosions of energy that have been detected across the sky but so far have defied explanation.

As with white dwarfs and the Chandrasekhar limit, neutron stars also have an upper limit on mass. Those exceedingly rare stars having more than 30 times the Sun's mass can produce cores containing 2.5 solar masses or more. For these objects, "implosion is compulsory," writes physicist Kip Thorne in his marvelous book, *Black Holes & Time Warps: Einstein's Outrageous Legacy*. "For stars of sufficiently large mass, neither the degeneracy pressure of electrons nor the nuclear force between neutrons can stop the implosion. Gravity overwhelms even the nuclear force."

The result is a black hole, a so-called singularity—essentially a point—with such intense gravity that not even light can escape. The boundary beyond which an object would need to move faster than light to escape a black hole's gravitational clutches is called the event horizon. Anything that ventures inside the event horizon effectively disappears from the known universe.

"The absolute maximum mass a neutron star can have is about two-and-a-half solar masses and some would say it may be even lower than that," explains Mario Livio, a theorist at the Space Telescope Science Institute. He adds that a star heavier than 30 solar masses or so "would start by trying to form a neutron star at the center. But then material will rain down and increase the mass beyond the point where the thing can support itself against gravity and it will collapse to form a black hole."

A star between 8 and 30 solar masses, then, will undergo a supernova explosion that results in a neutron star; more massive stars "almost certainly form black holes," Livio says. By definition, black holes cannot be seen. But their gravitational effects—unique signatures that can't be explained by other known phenomena—can be directly observed. In fact, the gravitational field of a black hole is no different from that of a normal star with equivalent mass.

Like a cosmic parasite, a white dwarf—in the middle of the disk at upper right—sucks material from its neighbor, a cool "failed star" that never achieved the sustained nuclear power production that defines true stars. Gas and dust streaming between the two contributes to an accretion disk around the white dwarf. The binary stars portrayed in this artwork are about one solar diameter apart; if real, they would orbit each other every four to six hours.

"Things only start to get bizarre when you really start approaching the horizon because it is then that tidal forces, or if you like, the curvature of space-time, starts to become very, very large," Livio says. "You would be stretched and torn apart, the atoms in your body torn apart..., time would start going slowly. But as long as you look at it from a distance, it would not look any different than the same mass accreting matter. The only thing is you would not see the central object."

Imagine a binary system made up of a black hole and a red giant. As the larger star blows off its outer layers, some of this material would spiral into the black hole, forming a flattened disk. Gas from the red giant would be accelerated to enormous velocities and heated to extreme temperatures. The accretion disk around the black hole would emit detectable ultraviolet radiation, x-rays, and gamma rays.

Livio adds that at least nine candidate stellar-mass black holes have been identified in binary-star systems. Relativity theory sets no upper limit on the mass of a black hole, and supermassive holes are believed to lurk in the cores of many galaxies, including the Milky Way. So far, about 40 such galactic black holes have been identified, by measuring the velocities of nearby stars.

"If they are not black holes, they are something even more bizarre," Livio says. The heavyweight record to date is held by an unimaginably huge black hole at the heart of galaxy M-87, in the constellation Virgo; its mass is equivalent to three billion Suns. It is not yet clear how such supermassive galactic black holes form. Regardless of the mechanics, though, black holes remain one of the hottest topics in astrophysics. And one of the most fun to contemplate.

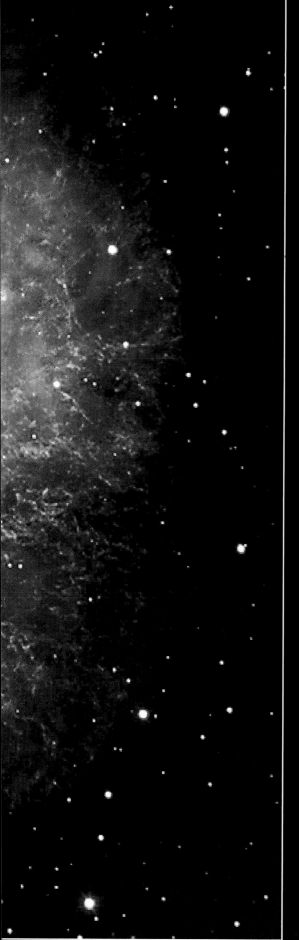

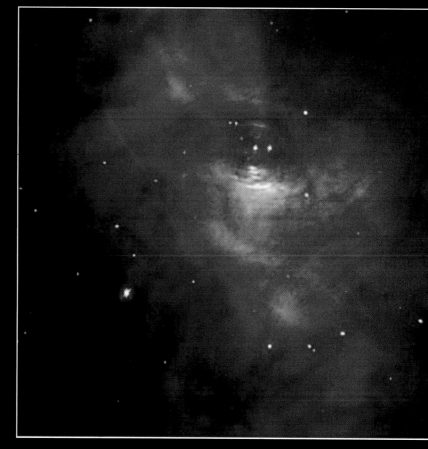

Today's magnificently variegated Crab nebula (left)
arose from a star that exploded into a supernova in
A.D. 1054 and was recorded by Chinese astrologers.
Too small to see in this ground-based image is its tiny
beating heart, a pulsar only six miles wide but more
massive than our Sun. A Hubble close-up (above)
reveals that the pulsar—leftmost star of the central
pair—is clearing an area as it spins 30 times a second
and sends particles hurtling out into the nebula.

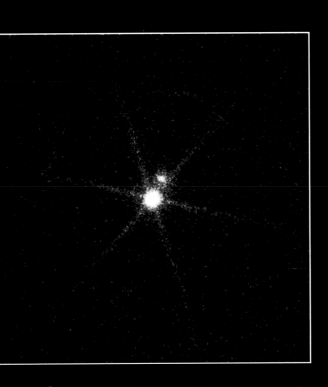

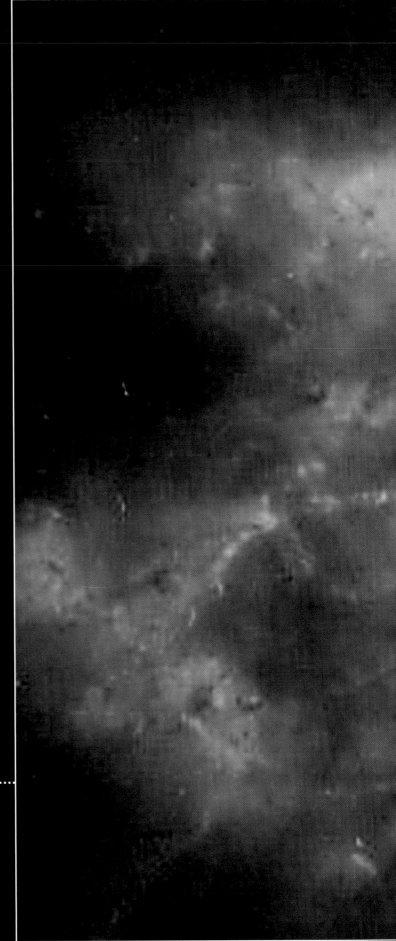

Brightest star in our night sky, Sirius (above) actually is a binary system: the white dwarf Sirius A orbited by the smaller, nearly obscured Sirius B. Astronomers long ago deduced that it was binary due to its slightly irregular path; astronomer Alvan Clark first detected Sirius B in 1862. The gravity on the surface of this white dwarf is 400,000 times that of Earth's.

White dwarf (right) lurks within the shroudlike remains of the planetary nebula NGC 2440. Planetary nebulae have nothing to do with planets but are the clouds that result from the death of a star the size of our Sun. This dwarf is so hot—perhaps 360,000°F— that it illuminates its surrounding nebula.

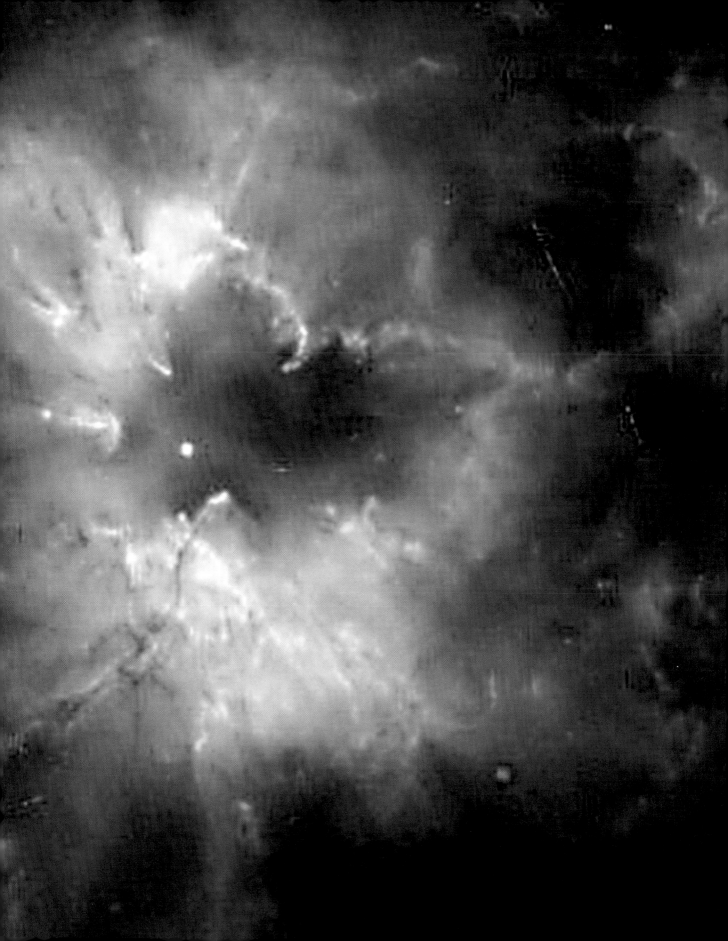

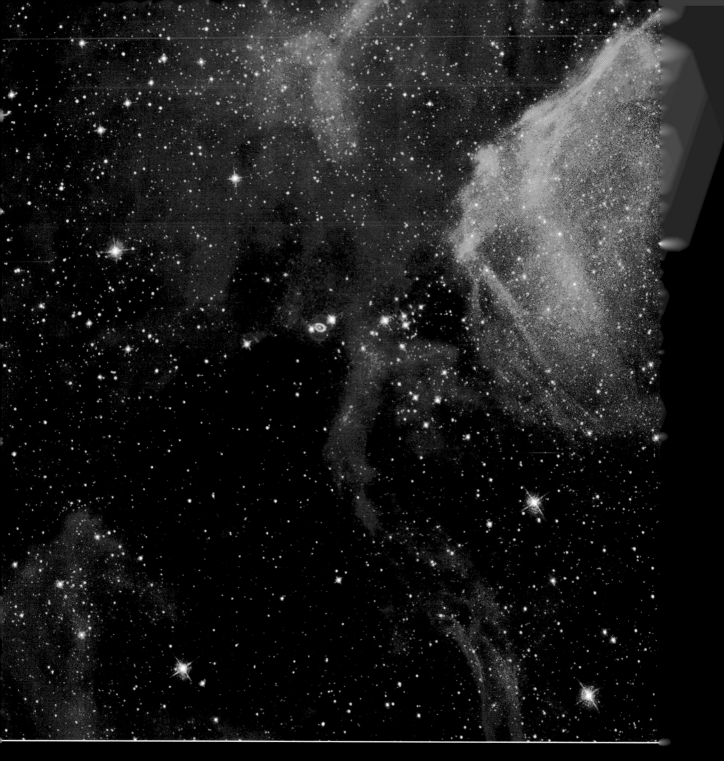

Glittering stars and diffuse clouds of gas create an ethereal backdrop for tiny but majestic supernova 1987A (above, at center); a close-up (right) reveals the complex structure of its glowing gas rings. Most stars live as red giants for a billion or so years, then quietly burn out and become white dwarfs. Larger ones—those at least eight times as massive as our Sun— explode into supernovae. Such stars first evolve into supergiants, producing different elements in their nuclear cores while losing part of their bulging envelopes. As their nuclear fires wane, they eventually surrender to their own gravity and, in a fraction of a second, collapse and blow apart. In the Middle Ages, supernovae sightings upset one tenet of Ptolemaic thinking: that the outermost sphere of the heavens, the realm of stars, was constant and unchanging.

Galactic cannibal, a massive black hole (above) hides at the center of a galaxy that is itself feeding on a smaller galaxy. This black hole, about ten million light-years from Earth in Centaurus A, sucks a disk of hot gas into its gravitational whirlpool. The disk is tilted in a different direction from the black hole's axis—like a wobbly wheel around an axle.

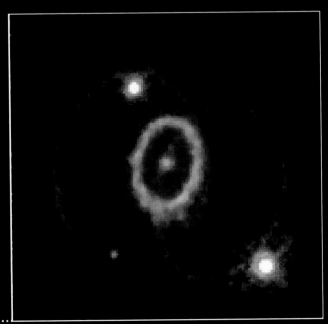

Searching for Other Worlds

Extrasolar Planets

When I was a boy, my older brother Bob and I would occasionally while away a summer evening under the stars, reclining on lawn chairs and letting our peripheral vision catch the brief streaks left by shooting stars. Bob introduced me to science fiction at an early age and, resting under the dome of the night sky, we would marvel at the majesty of the universe, debate our own insignificance, and speculate about how many of the stars we saw sprinkled across the heavens might harbor solar systems or even alien civilizations. It never occurred to either of us that, in a galaxy of 200 billion suns, our solar system might be the only one.

We were not alone in that assumption, of course. But given the enormous distances between the stars and the unavoidable fact that a typical star would outshine the reflected light from any of its planets by about a billion

In the ongoing search for solar systems similar to our own, astronomers have so far discovered 58 planets orbiting other stars. This artwork depicts a pair of such extrasolar planets performing a stately two-step around their star, the red giant Gliese 876, as seen from a hypothetical moon.

to one, Bob and I could not imagine how one might go about proving the reality of those presumed worlds— barring direct contact with alien beings, of course, or the invention of a faster-than-light warp drive. And so, another round of animated discussion would begin and run its course.

As it turns out, our imaginations were limited by our lack of knowledge. In 1968, about the time we were looking for meteors and debating the reality of aliens, Swarthmore College astronomer Peter Van de Kamp announced the results of a painstaking reanalysis of data that, he said, indicated the presence of a Jupiter-class planet in orbit around a star just six light-years from Earth. Van de Kamp had spent decades carefully monitoring the target star against the background of interstellar space, measuring a tiny back-and-forth wobble he believed was caused by the gravity of some unseen companion tugging on the star. Others using similar techniques had made similar claims for other stars during the early 1940s, but their observations could not be confirmed. Van de Kamp cited evidence for an extrasolar companion of some sort in 1943. Twenty years later, he felt confident enough to claim it was a planet, and five years after that, he bolstered his earlier conclusion with additional analysis.

But other astronomers later challenged his results and, in 1973, proved to the satisfaction of the astronomical community that no such planets existed. There matters stood until 1991, when Alex Wolszczan, using the great radio telescope at Arecibo, Puerto Rico, noticed unexpected changes in the timing of signals from a spinning pulsar. He ultimately concluded that the changes were caused by the gravitational effects of three planets orbiting the pulsar, a discovery that made headlines around the world. But because pulsars— collapsed stars—are strange beasts, it was not immediately clear what his discovery meant about the likelihood of planets circling more Sun-like stars.

In 1995, astronomers Michel Mayor and Didier Queloz used spectroscopic techniques to discover a Jupiter-class planet orbiting within a few million miles of the star 51 Pegasi, far closer than current theories

of planetary evolution predicted. Shortly thereafter, a team lead by Geoffrey Marcy and R. Paul Butler found massive planets around two more stars. In another major surprise, one of these planets followed a highly elliptical path, in stark contrast to the more circular orbits of planets in our own solar system.

Since then, finding extrasolar planets has become one of modern astronomy's hottest fields. By June 2001, a database maintained by the Paris Observatory listed 58 confirmed extrasolar planets orbiting main sequence stars, 2 pulsar-type planetary systems, 3 protoplanetary disklike structures, and another 14 "unconfirmed, doubtful, unpublished, or invalidated objects." No detectable planets were found around 21 other stars that had been examined to that point using similar techniques.

Of the distant solar systems that have been confirmed, astronomers have yet to find one very similar to ours, in which Jupiter-like planets orbit parent stars in nearly circular paths and at relatively large distances. Instead, astronomers have been astonished to find huge Jupiter-size planets orbiting fairly near their suns. This is most likely a result of the methods used, not an indication of some universal law governing planetary formation. With current techniques, the nearer a planet is to its primary star, the easier it is to detect. Many astronomers believe the unusual number of "hot Jupiters"—large planets that travel so close to their sun that they're heated to several thousand degrees— discovered to date is an artifact; as instrumentation improves, more Sun-like solar systems should turn up. But making predictions in a field as dynamic as this is a risky business and no one really knows what will be discovered in the years ahead.

Planetary theorist Alan Boss, at the Carnegie Institution in Washington, D.C., says, "We're really jazzed up and excited" about the discoveries of hot Jupiters. "Because it means there's something there. Even looking where no one thought there would be planetary systems we're finding them like mad; they're all over the place. We're slowly lowering the veil on the area of what we can see; things are getting more and more interesting.... We're hoping that when the veil is dropped completely, we'll truly be astounded by the bonanza of planetary systems we'll have."

The basic method for indirectly detecting extrasolar planets was developed more than three centuries ago by Isaac Newton, who put gravitational interactions on a mathematical footing, and the German mathematician Johannes Kepler, who formulated a set of laws governing orbital motion. The velocity of a planet, moon, or any other orbiting body depends only on the radius of the orbit. The closer a body is to its "sun," the faster it travels. Thus a space shuttle orbiting Earth at an altitude of 200 miles takes about 90 minutes to complete one orbit, while a communications satellite orbiting 22,300 miles above the Equator takes 24 hours—and appears to hang stationary in the sky as the Earth turns. (A 24-hour orbit is extremely handy because, as Arthur C. Clarke first realized in the mid-1940s, it allows the use of stationary ground antennas, which in turn vastly simplifies the technology required for globe-spanning communications networks.)

The relationship between an object's velocity and its orbital radius is easily demonstrated within our own solar system. Mercury needs only 88 days to circle the Sun at an average distance of 36 million miles, while Pluto—3.7 billion miles out—takes 248 years to orbit the Sun once. Interestingly, the orbiting body's mass does not significantly affect either its orbital period or its velocity.

Gravity also varies with distance. Whether or not a falling apple triggered the realization, Isaac Newton discovered that the gravitational force between two bodies is inversely proportional to the square of their distance from each other. Thus a planet orbiting four units from its sun will experience one-sixteenth of the gravitational attraction that a planet only one unit away experiences. Consider the Sun and Jupiter, ignoring for the moment all other components of our solar system. The Sun's gravity keeps Jupiter in an 11.9-year orbit, exactly as predicted by Newton. But Jupiter's gravity also tugs on the Sun, pulling it ever so slightly toward the planet. In fact, the Sun and Jupiter actually orbit

a common center of mass, and the amount of their mutual tugging is directly proportional to the masses involved. Because the Sun outweighs Jupiter about a thousand to one, the common center of mass is one thousand times closer to the Sun, a point about 31,000 miles beyond the Sun's upper atmosphere. Viewed directly above the plane of the solar system, the Sun moves around this point as Jupiter sails along in its orbit. The picture grows more complicated, of course, when one adds the effects of the other planets, but the basic result is the same: The Sun wobbles ever so slightly as the planets wheel about. We can detect it, because we're so close. Viewed from 33 light-years away, however—just around the corner as interstellar distances go—the Sun's motion due to Jupiter's gravitational influence would be comparable to the size of a dime seen from 1,000 miles away. How to detect so subtle a motion from afar?

One way is to measure the actual movement of a star in space, the same technique Van de Kamp used. Another is to measure the intensity of light from a broad population of stars. When an extrasolar planet moves in front of its star as viewed from Earth, the star's light will dim slightly and then brighten. By detecting and measuring such periodic fluctuations in intensity, astronomers can determine the nature of the objects that cause the dimming.

NASA is considering developing a spacecraft it calls Kepler, which would use just such a technique to monitor nearly 200,000 stars for signs of extrasolar planets. Based on pure statistics, Kepler would be expected to detect more than 600 Earth-size planets and another 1,700 or so Jupiter-class planets, assuming they orbit relatively close to their parent stars. In addition, NASA and the European Space Agency are contemplating even more sophisticated spacecraft that could image any Earth-size planets found around relatively nearby stars.

So far, the most productive technique for finding extrasolar planets uses ultra-sensitive spectroscopes to measure a star's radial velocity around the star-planet system's center of mass. Just as the pitch of a siren changes as it approaches and then recedes, so does the wavelength of a star's light change as it approaches or moves away from the observer. Such changes in stellar wavelength are on the order of one part in a hundred million, but they are detectable.

"You're looking at the star and you're getting a lot of photons from the star and you're looking for absorption lines in the stellar spectrum...that identify features in the star's atmosphere," Alan Boss says. "You watch those lines shift in wavelength as the star moves back and forth along the line of sight to you, because the star is going around the center of mass of the star-planet system."

With enough observations, radial velocity data can provide the planet's minimum mass, its distance from the parent star, and its orbital time. Other data are needed to determine characteristics such as the tilt of the planet's orbit relative to the star's equatorial plane.

"It's really powerful if you have both," Boss says. "Because in reality, any of these solutions is going to be sort of based on slightly noisy data."

Current technology clearly favors the discovery of massive planets that orbit relatively near their stars. Whether such relationships ultimately prove to be common or uncommon is not yet known, but just realizing that hot Jupiters exist has been a revelation for planetary theorists trying to figure out just how solar systems evolve. It is not surprising that a planet's orbit might change over time; theorists have considered that possibility since the 1980s, and recent data from the Galileo probe indicate that Jupiter actually may have taken shape much farther from the Sun than it is now, and only later moved to its present location. What is surprising about the hot Jupiters discovered so far is how close many of them are to their suns. It's possible that gravitational interactions among other massive, nearby planets—planets yet to be directly detected—play a role.

"But," says Boss, "you really have to play a lot of billiard games in order to manage to sink the shot that puts something close to a hot Jupiter's orbit. You see a lot of hot Jupiters, so you need to have a pretty efficient

An Earth-like moon—complete with water and an atmosphere—orbits a Jupiter-size planet in this artist's rendering. Although the moon is imaginary, the planet is known to orbit the star HD177830, located 192 light-years away in the constellation Vulpecula. Generally, astronomers search for extrasolar planets using indirect methods, such as seeking tiny wobbles in a star's path that might indicate a gravitational partner.

mechanism. Gravitational billiards, I think, is a long-shot candidate." A more likely scenario, he feels, involves gravitational interactions between a growing Jupiter-class planet and the solar nebula that spawns it. Such interactions can kick a planet into a higher orbit or even eject it from its solar system. More often, the end result is a loss of angular momentum, causing the planet to migrate closer to its sun.

But all this begs the question of why Jupiter and Saturn are in the outer regions of our solar system, not closer to the Sun. Given Earth's history, it's not merely an academic question. If hot Jupiters are the rule rather than the exception, life in this universe could be rare indeed. Again, Boss explains: "We still hope we're going to find solar-system analogs that look something like our solar system...with a nice hefty gas-giant planet out on a long-period orbit that can protect an Earth-like planet in a short-period orbit. Because Jupiter's important for keeping the comets off our back and not having us go the way of the dinosaurs too often."

Assuming the current pace of technology continues, Boss expects ground-based instruments to begin detecting such "solar-system analogs" in the next five years, as astronomers follow Jupiter-class planets through longer-period orbits. Space-based planet finders will begin coming on-line around the end of the decade. Only time will tell whether solar systems like ours are common or rare. But based on past experience, it's a safe bet the universe will turn out to be even more surprising than we currently imagine.

Galaxies and the Distant Universe

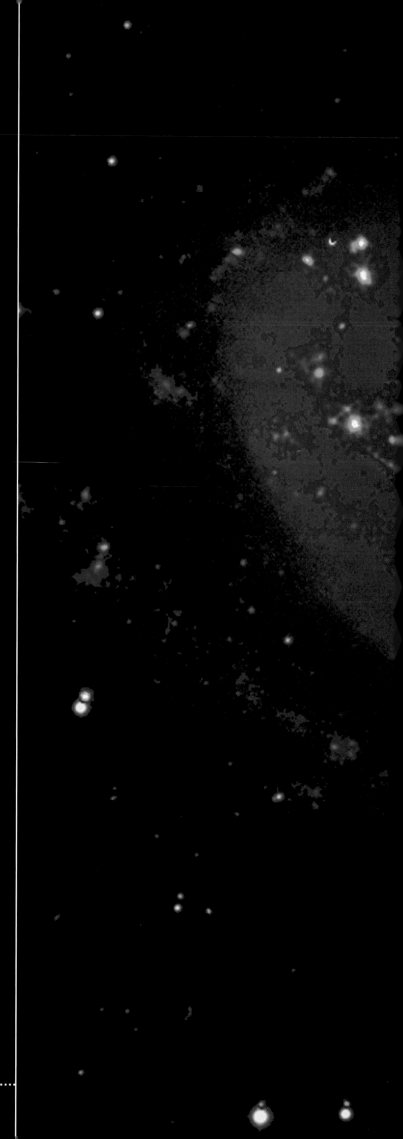

The target was a tiny patch of empty sky near the Big Dipper, a cosmic keyhole no larger than a grain of rice held at arm's length. Test photos showed nothing of any significance, no bright intervening stars, no galaxies, no obscuring clouds of dust or gas. Just empty space, an unobstructed view out of the disk of the cluttered Milky Way and into the depths of the expanding universe.

On December 18, 1995, the Hubble Space Telescope obediently wheeled about, locked onto nearby guide stars and began snapping pictures of this seemingly empty patch of sky. Lots of pictures. For the next ten days, Hubble took one after another in ultraviolet, blue, red, and infrared light, slowly building up a digital image that reached further and further back in time, collecting photons emitted when the universe was a fraction of its present age.

Like all spiral galaxies, the elderly giant M81 features a bulging middle and a pronounced pinwheel shape. Many of the smudges of light we see in the heavens actually are galaxies of one sort or another; even the smallest ones contain millions of stars. About seven million light-years from Earth, M81 lies within the constellations Ursa Major and the Big Dipper; it can be seen with good binoculars.

Later, as the 342 individual images were processed and combined into a single color photo, astronomers at the Space Telescope Science Institute, in Baltimore, were stunned by what they saw: Some 1,500 discernible galaxies or fragments of galaxies, sprinkled across the image like uncut gemstones, astonishing in their diversity. All in an "empty" speck of the sky.

That picture is known as the Hubble Deep Field and, for sheer intellectual vertigo, it's in a class by itself. Consider it an astronomical core sample, a glimpse through the depths of space and time to a point 12 or 13 billion years ago, only one billion years after the Big Bang created the known universe. In this single image, astronomers see the full range of galactic evolution, from fragmented assemblies of hot young stars near the dawn of the universe to the exquisite spirals and huge elliptical galaxies that evolved later.

The Hubble Deep Field appears to confirm the predictions of hierarchical galaxy formation, a theory that views galaxies as the result of an initial phase of gravitational collapse followed by mergers and collisions that triggered repeated bouts of star formation and mass redistribution. The surprising brightness of the youngest galactic fragments in the photograph is consistent with the idea that galaxies grow through mergers, explains University of Wisconsin astronomer John Gallagher, an expert on galactic evolution.

"When two good-size hunks of protogalaxy or youthful galaxy hit each other, you get a violent explosion of star formation and that kind of starburst produces high brightness," he says.

In a narrow but greatly magnified "keyhole" view that reaches nearly to the universe's visible horizon, this image includes about one-fourth of the Hubble Deep Field, which displays some of the most distant galaxies yet seen. Many hundreds of galaxies crowd this frame; light from the faintest of them has taken 11 billion years to reach Earth. The area shown represents but a flyspeck of the dome of our night sky.

At the beginning of the 20th century, astronomers were unsure whether other galaxies even existed outside the Milky Way. Many nebulae were known and many were shaped like spirals. British astronomer William Herschel, who discovered Uranus in 1781 and was one of the first to systematically map the Milky Way, concluded that so-called spiral nebulae were in fact separate galaxies. Others believed Earth's home galaxy was the only such structure in the universe and that all nebulae and star clusters visible at night were members of this single system. The issue would not be resolved until 1924, when American astronomer Edwin Hubble, the man for whom the Hubble Space Telescope is named, identified Cepheid variable stars in what was then known as the spiral nebula in Andromeda.

Cepheid variables are yellow giant stars that regularly brighten and dim, or pulsate, over time. The time required for a Cepheid variable to do this is proportional to the overall luminosity of the star. The brighter the star, the longer the time between pulses. Hubble measured the apparent brightness of Cepheids in the

Andromeda star swarm and used a period/luminosity scale that was based on the absolute magnitudes of nearby Cepheids to determine their true brightness. Since the intensity of starlight decreases by the square of the star's distance from the observer, Hubble then was able to calculate how far away the Andromeda Cepheids must be. Although his rough estimates were off by a considerable margin, they were close enough to establish beyond any doubt that "spiral nebulae" like the one in Andromeda were, in fact, separate galaxies far beyond the Milky Way.

Hubble went on to develop a sort of taxonomy for galaxies, dividing them into two broad classes: spirals (like the Milky Way) and ellipticals (bloblike affairs lacking any arms). Both types were split into various subcategories. About 15 percent of known galaxies fall outside this conceptual framework. Disk galaxies, for example, exhibit a flattened shape but no spiral structure. Irregular Type I galaxies have some spiral features within a generally disorganized appearance. Irregular Type IIs include galaxies that have been deformed by collisions with other galaxies.

Hubble arranged all the major galaxy types into a graph that resembles a tuning fork. Spherical galaxies appear at the base of the fork's stem; ellipticals occur along the stem, growing increasingly elongated as they approach the disk-shaped galaxies, which show up at the base of the fork's two arms. Spiral galaxies populate one arm, while barred spirals—spirals displaying a thickening or elongation across their center—occupy the other. Initially, astronomers believed this arrange-

ment represented an evolutionary path, that galaxies began as sphericals and evolved into flattened-out spirals and barred spirals. But soon they realized the tuning-fork model could not be right, because galaxies at the end of the "sequence" had stars that proved to be just as old as those in the supposedly older galaxies.

Today, theories of galactic evolution hinge on the presence of an unseen "dark matter" believed to make up 90 percent of the mass in the universe. Although dark matter has never been directly detected, astronomers have little doubt it exists.

As far back as the early 1930s, Sinclair Smith and Fritz Zwicky, the eccentric genius who predicted the existence of neutron stars, measured the velocities of galaxies in the Coma cluster, a spherical swarm of more than 1,000 bright galaxies 300 million light-years from Earth. They then measured the brightness of the galaxies in the cluster and used that data to determine their combined mass. To everyone's surprise, their calculations showed the cluster would need 10 times its observed mass in order to gravitationally rein in member galaxies. Based solely on that observable mass, the galaxies would have flown apart long ago.

Fifty years later, Vera Rubin and Kent Ford, of the Carnegie Institution, in Washington, measured the rotational velocities of a wide assortment of galaxies. If those velocities depended only on the combined mass of the visible stars in each galaxy, those stars farther from the galactic core should have lower velocities than those closer in. But that is not what Rubin and Ford observed.

"One overwhelming conclusion emerges from our observations," Rubin wrote in a 1983 article for *Scientific American*. "Virtually all the rotation curves are either flat or rising out to the visible limits of the galaxy. There are no extensive regions where the velocities fall off with distance from the center, as would be predicted if mass were centrally concentrated. The conclusion is inescapable: Mass, unlike luminosity, is not concentrated near the center of spiral galaxies. Thus the light distribution in a galaxy is not at all a guide to mass distribution."

The most likely explanation, she wrote, is that each galaxy is embedded "in a spherical 'halo' of matter that extends well beyond the visible limits of the galactic disk. The gravitational attraction of this unseen mass keeps the orbital velocities of the galaxies from decreasing with distance from the galactic center."

What this unseen dark matter might be remains an open question. It could be huge numbers of small, dim stars, brown dwarfs, or other objects known collectively as massive compact halo objects, or MACHOs. Others believe it may consist of what are called WIMPs—weakly interacting massive particles—which are too heavy to be produced by today's particle accelerators. Whatever dark matter is, its collective gravitational effects are enormous; dark matter has played a role in the formation and evolution of galaxies and galactic clusters. To get a sense of what might be going on, we must keep the broad outlines of Big Bang cosmology in mind.

By spectroscopically measuring the velocities of a sampling of galaxies, Hubble discovered that the universe is expanding, that galaxies are flying apart at speeds proportional to their distance from each other. That means that at some time in the distant past, all the mass in the universe must have been concentrated in a single point. And in the distant future? Is it reasonable to assume that, ever since the Big Bang, the combined gravity of everything in the universe, both dark matter and normal, has been acting to slow the cosmic expansion? If enough matter exists, expansion might be stopped or even reversed.

But astronomers have not found nearly enough matter to halt the cosmological expansion. In fact, observations of distant type Ia supernovae, which serve as cosmic mile-markers over greater distances than those spanned by Cepheid variables, indicate that expansion actually may be accelerating. This astonishing conclusion implies the existence of some sort of "dark energy" that, like dark matter, pervades the universe yet cannot be directly detected.

While the jury is still out on whether the universe is accelerating, there is general agreement today on what

happened at the other end of time's arrow. Modern astronomers believe the known universe—matter, time, and space—exploded into existence 12 to 14 billion years ago. The fireball of matter and energy rushing outward from this inconceivable blast did not expand into some space-time framework that already existed. Rather, space itself expanded, carrying matter and energy along. By about 300,000 years after the Big Bang, the universe cooled to the point where light could move through space and the cosmos had become transparent. The light from that era is still around today—in the form of stretched-out microwave radiation that corresponds to a background temperature of about 2.7 degrees above absolute zero. This is nothing less than the faint afterglow of the Big Bang itself.

It now appears that the expanding fireball of the Big Bang was not perfectly uniform and featureless. In less than a billion years or so, from the time the universe became transparent to the time when galaxy fragments began showing up in the Hubble Deep Field, regions of higher density gave birth to stars, which then organized themselves into galaxies and clusters of galaxies.

"Star formation can occur rapidly, in less than a few million years," says John Gallagher. "So once a galaxy gets started, it can produce stars pretty quickly." He paints an intriguing picture of an infant universe in which areas of slightly higher density, dominated by dark matter, lead to the formation of galaxies and vast clusters embedded in halos of the unseen matter.

"If you make a little density perturbation in space, which we think comes to us from earlier in the universe, then what happens is that region expands more slowly than its surroundings. And because it's expanding more slowly, it stays denser than its surroundings. So as this universe tends to expand, these things are kind of getting left behind, they're becoming denser and denser relative to their surroundings."

Thus the gravity in these denser regions—generated primarily by the dark matter that dominates the universe—becomes stronger and stronger compared to the forces driving the expansion of the universe

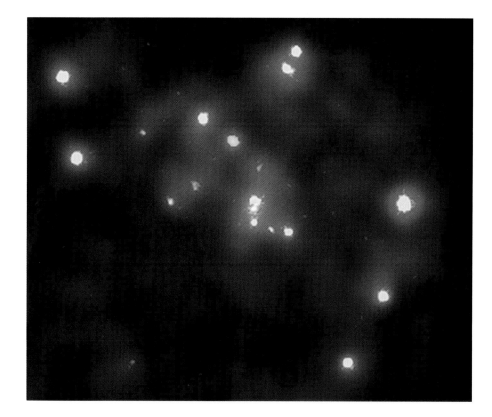

Gigantic black hole with the mass of 30 million Suns—shown here as a central blue dot—lurks at the core of the Andromeda galaxy, about two million light-years away. Matter funneling toward the black hole produces x-rays that enable its detection. A spiral galaxy similar in shape and size to the Milky Way, Andromeda contains some 200 billion stars.

itself. Eventually, these objects stop expanding when their own gravity takes over, "fall back on themselves, and form objects that are bound together by gravity," Gallagher explains. "That's the basic mechanism for galaxy formation. I think we're pretty clear on that. The trick is how to make the models produce objects like real galaxies; we really aren't able to do this yet."

The most popular current model is one in which the first objects to "turn around," or stop taking part in the cosmological expansion of the universe, are relatively small structures. "These small objects," says Gallagher, "kind of drop out in clouds and they then conglomerate together because they're near each other. This is called hierarchical galaxy formation. You make small pieces that stop expanding; the small pieces then combine together to make larger galaxies."

Summarizing, he says, "You tend to make a bunch of small things out of the dark matter, these small things garner the gas, the gas turns into stars, the small things smash into each other and become bound into bigger units. The big things eat all the small things, and you end up with most of the matter in the big things and a few small things left over."

Galaxies tend to belong to vast gravitationally bound groups, or clusters, which tend to be associated with even more enormous structures called superclusters. The Milky Way, the Andromeda galaxy, and other members of our local group are part of one such supercluster that also includes the Virgo and Fornax galaxy clusters, among others. Its center of mass appears to be in or near the Virgo cluster, some 50 to 60 million light-years away.

One major unknown in the hierarchical evolution of galaxies is the distribution of dark matter on small scales during the initial stages of galaxy formation. And how did stars form in the first place? Astronomers believe the gravitationally driven collapse of huge clouds of gas inevitably leads to star formation, but the details are not known.

Today's galaxies, however, are not the relatively stable, static structures of Hubble's day. We now know they undergo major changes over their lifetimes, just as generations of stars come and go, redistributing normal matter over time scales that are beyond human comprehension. More profound changes occur as galaxies merge or collide, triggering waves of explosive star birth and perhaps causing those bright galaxy fragments that are visible in the Hubble Deep Field. Over time, spirals and disks tend to become more spheroidal with more tightly concentrated regions of normal matter.

"For a long time, the debate was 'Is it nature, how the thing was formed, or nurture, how it grew up,'" says Gallagher. "And the answer seems to be 'Yes.'"

Fighting a losing battle, a dusty spiral galaxy—the dark band—gradually finds itself being absorbed by a giant elliptical galaxy, Centaurus A (left). Some 13 million light-years from Earth, Centaurus A emits about 1,000 times more radio energy than our own galaxy. Three exposures—red, blue, and green, combined for this image—reveal many tiny red star-forming regions and clumps of young blue stars recently formed from them. The black hole hidden here may contain the material of a billion of our Suns.

Following Pages: With a glancing blow, two spiral galaxies in the direction of the constellation Canis Major pass each other by. The larger, more massive one at left emits strong tidal forces that distort the smaller galaxy, flinging stars and gas into titanic streamers that stretch perhaps 100,000 light-years to the right. Circling its neighbor in a counterclockwise direction, the small galaxy made its closest approach about 40 million years ago; gravity eventually will bring it round again. In the even more distant future—billions of years from now—the two will likely merge and become one massive galaxy. Our Milky Way probably grew in similar fashion.

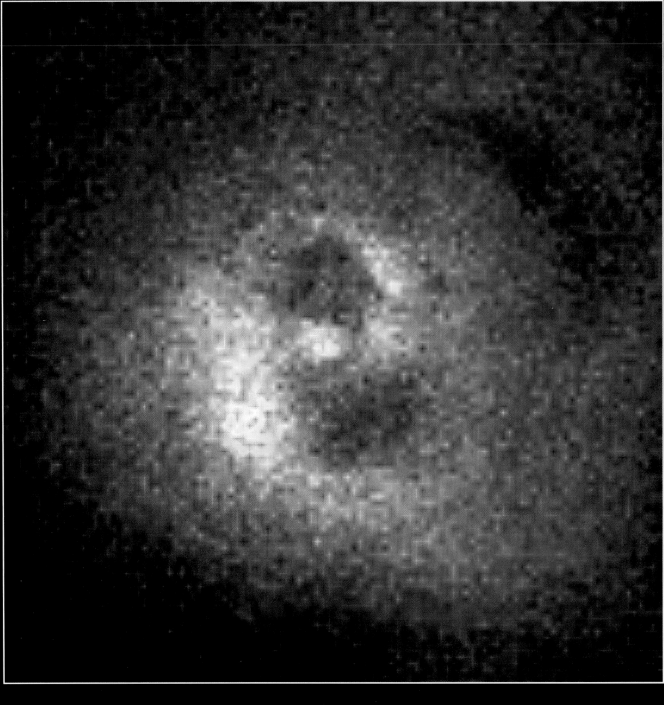

Ravenous supergalaxy Perseus A (above) grows larger by stealing from others. This historic image from the Chandra X-ray Observatory allows astronomers to see an x-ray shadow cast by a small galaxy as its gas is stripped away by a larger one. Perseus A, in the center of the image, has accumulated thousands of galaxies and hundreds of billions of stars; it is one of the most massive aggregations in the known universe. The small dark patch at about two o'clock represents x-rays from the defeated galaxy, a collection of some 20 billion stars that is falling into Perseus.

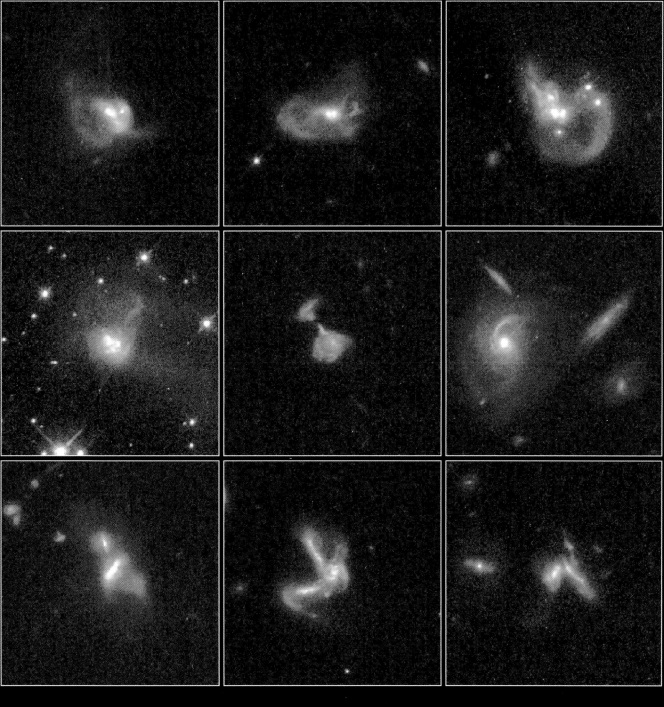

Like galactic bumper cars, galaxies collide; unlike
them, they often merge. These Hubble images show
three-, four-, and even five-galaxy pileups. Astronomers studying ultraluminous infrared galaxies
(ULIRGs) found these and at least 15 other examples
of multi-galaxy collisions, which create tangled
clumps of matter and trigger bursts of new stars.
The nuclei of several galaxies appear in the upper
right photo; at bottom center a three-galaxy scuffle
rips streamers of stars from their homes. Similar

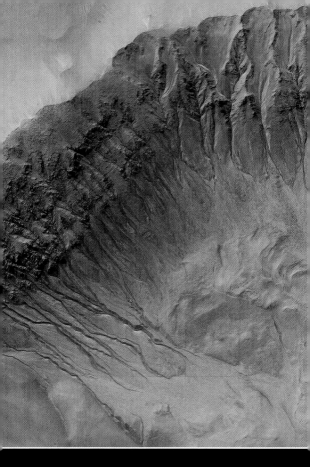

Its form ever more familiar to us, Mars will be one of tomorrow's astronomical classrooms. Here, gullies and slides imply a surface etched by Earth-like forces,

PART II

THE NEXT DECADE

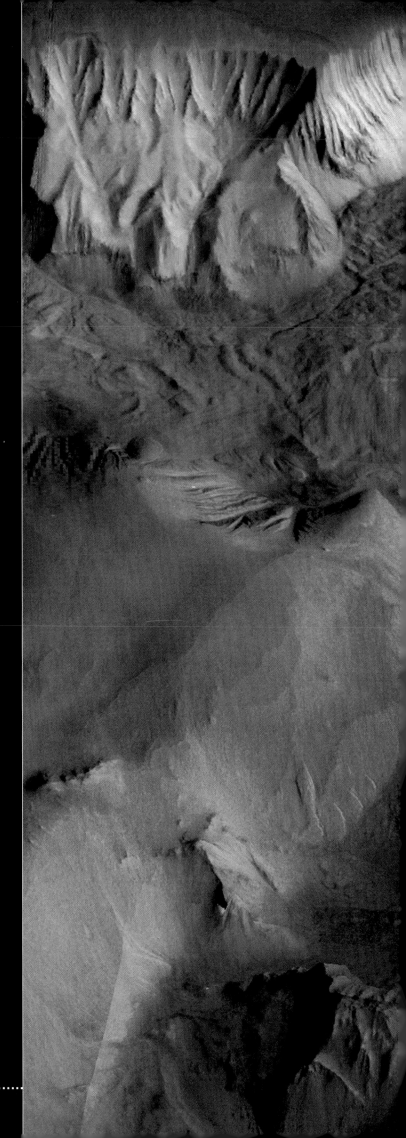

RETURNING TO MARS

S treaking through space at some 13,000 miles per hour, a small spacecraft will slam into the atmosphere of Mars in early 2004, following a six-and-a-half-month voyage from Earth. In just two minutes it will slow to less than 230 mph and put out a parachute. Several minutes later, cocooned in protective air bags, the probe will hit the surface between 25 and 50 mph, bouncing and tumbling up to a half mile before finally rolling to a stop.

In short order, the air bags will deflate, the spacecraft will open like a blossom, and a squat solar-powered mobile laboratory will slowly wheel away, beginning NASA's first visit to the surface of the red planet since Mars Pathfinder and its Sojourner rover captured Earth's imagination nearly seven years earlier, in 1997. But unlike the one-shot Pathfinder, this robotic geologist will have an identical twin, scheduled to touch down at a different

Slumping landslides texture the Martian surface, which NASA plans to explore in the coming decade with robotic landers that will assess its structure, chemical composition, geologic history, and climate. The success of such missions could lead the way, eventually, to human exploration of the red planet.

Martian site a month later. Both probes are part of a new campaign that began with the launch of the Mars Odyssey orbiter in April 2001 and aims to study Mars as an integrated system, mapping its structure and composition while determining its geologic history and climatology through a sharply focused and technologically adept set of missions. In so doing, NASA hopes to find out what happened to the water that once shaped the Martian surface and whether life ever managed to evolve on the fourth planet from the Sun.

"We're pretty certain that water [once] flowed profusely on Mars, carving out the channels we see today," Ed Weiler, NASA's associate administrator for Space Science, told me before Odyssey's launch. "In fact, recent results from the Mars Global Surveyor tell us that water not only flowed billions of years ago, but it might have flowed a million years ago. And we can't rule out that it flowed a few days ago."

On Earth, water, organic compounds, and energy are the basic ingredients of life. Virtually anywhere we find all three, we also find life, even at the extremes of Earth's environment. All three almost certainly existed in Mars's distant past and may still exist there today.

"Finding life on Mars, whether it's past or present, is a major goal of the Mars program," Weiler said. "It is not the only goal. Along the way we want to understand Mars as a planet, its geology, its climatology, its surface composition, et cetera, which will help us understand our Earth in the past and in the future."

This new view of Mars as a still-active world that, in theory at least, could harbor some form of microscopic life even today, contrasts with the Mars of the 1960s and '70s, when the first robotic probes beamed back grainy black-and-white images of a Moonlike, heavily cratered surface. Many researchers concluded Mars was "bizarre, enigmatic, perhaps akin to the most exotic polar deserts here on Earth," recalls James Garvin, NASA's chief scientist for Mars Exploration. "And that's what we've now come to call the old Mars."

Life on Mars? Humans have pondered the possibility for generations. In the late 1800s astronomer Percival Lowell described what he saw as the planet's complex system of engineered canals; novelists Edgar Rice Burroughs and H.G. Wells planted fictitious civilizations there. Today's biologists find similarities between Earth's magnetotactic bacteria (top) and magnetite crystals found in a Martian meteorite that landed in Antarctica (above).

Old Mars had "shut off the engine of activity," explains Garvin. "Yes, there may have been water, but it was lurking in the past. We couldn't agree on how much." But by the late 1990s a new picture began to emerge, based primarily on data from the Mars Global Surveyor, which revealed snakelike channels carved in the sides of high-latitude craters—"gullying events" that appear to have been caused by the flow of water from beneath the surface.

Asks Garvin, "Could it be something else? Of course. Do we know all the answers? No. But I think no one would argue the landscapes weren't made by a fluid. What fluid, of course, will be debated. Many of us think water is the simplest answer."

In late 2000, additional images showed layers of what seem to be sedimentary rock in craters and other places where water might have pooled in the past. The presence of sedimentary rocks on Mars, if confirmed, would imply the long-standing presence of water, critical to the evolution of life. Learning more about the red planet has been difficult, however. Of 30 U.S., Russian, and Japanese missions planned for Mars between 1960 and the end of 1999, 5 produced little or no useful data and 16 failed outright. Loss of the $1-billion Mars Observer in 1993 prompted NASA to adopt a "faster, better, cheaper" approach to exploration that called for launching low-cost landers and orbiters every two years, starting with Global Surveyor and culminating in a robotic sample-return mission planned for 2005.

In late 1999, the Mars Climate Orbiter and Mars Polar Lander were lost within three months of each other. NASA quickly canceled plans for a 2001 lander and launched a major reassessment. At the same time, Global Surveyor's ongoing flood of data began leading NASA planners to a new "follow the water" strategy. Scott Hubbard, called in to oversee the new Mars program, settled on a conservative approach that minimized reliance on less mature technologies. Also, NASA decided to spread out the orbiter and lander missions, allowing more time to recover from problems and respond to potential scientific discoveries.

"We're going in phases," Hubbard explained. "And those phases are first, reconnaissance, followed by surface measurements, and then ultimately by a scientifically very well selected sample that we bring back to Earth." Toward that goal, NASA anticipates cooperation with France, Italy, Japan, the European Space Agency (ESA), and possibly Russia. ESA, for example, is mounting an ambitious $170-million mission dubbed Mars Express. Scheduled to arrive in December 2003, it features an orbiter that will drop a small lander to the surface. While the orbiter performs two years of remote sensing and photo mapping, the stationary lander will scoop up soil samples and look for signs of biological activity, past or present.

NASA's $297-million Mars Odyssey orbiter features a camera system capable of operating in visible and infrared wavelengths, a radiation monitor, and a gamma-ray spectrometer designed to measure the chemical composition of the upper few inches of the planet's soil and the amount of hydrogen that might be present in subsurface ice. The satellite's imaging system will look for surface hot spots during the Martian night, which might indicate active hydrothermal systems.

Data from Odyssey also will help scientists narrow the list of promising landing sites for the two rovers scheduled to launch in 2003 and arrive in 2004.

"The '03 landers are robot geologists," says Firouz Naderi, Mars program manager at the Jet Propulsion Laboratory. They will basically scratch the surface of rocks to expose pristine material, then use onboard imaging microscopes and spectrometers to analyze their composition. If all goes well on these voyages, a photographic satellite called the Mars Reconnaissance Orbiter (MRO) will be launched in 2005 to study the Martian atmosphere and map the planet in great detail, resolving surface features as small as two feet across. Hubbard says this data, combined with information gleaned from Odyssey, Global Surveyor, and the 2003 landers, will help researchers select "the best place or places where water was for a long time, the climate was probably warm and hospitable, the layered terrain looks like it might be sedimentary, and so forth."

In 2007 NASA hopes to launch a high-tech "smart rover" that will be capable of traveling seven miles or more across the Martian surface. Details are not yet finalized, but Garvin says the lander almost certainly will be able to gather soil samples from six feet or more below the surface, while the rover may have an electron microscope or even a special spectrometer that can identify organic compounds in those samples.

NASA plans yet another orbiter in 2009, followed in 2011 by its most ambitious and expensive mission to date, a technologically daunting flight to robotically collect and bring back to Earth about two pounds of rock and soil for laboratory analysis. This likely will involve two spacecraft: A lander with a smart rover (to collect samples) and a rocket (to boost the samples into orbit), and an orbiter to receive the sample cache and return it to Earth. The price tag may be as high as $2 billion, but the scientific payoff could be huge.

"If the first real Earth-like planet we look at shows there is evidence of having had a past biology, even a limited one that failed to sustain itself, my God," says Garvin, "that just opens the door to the possibilities in the solar system and the rest of the universe."

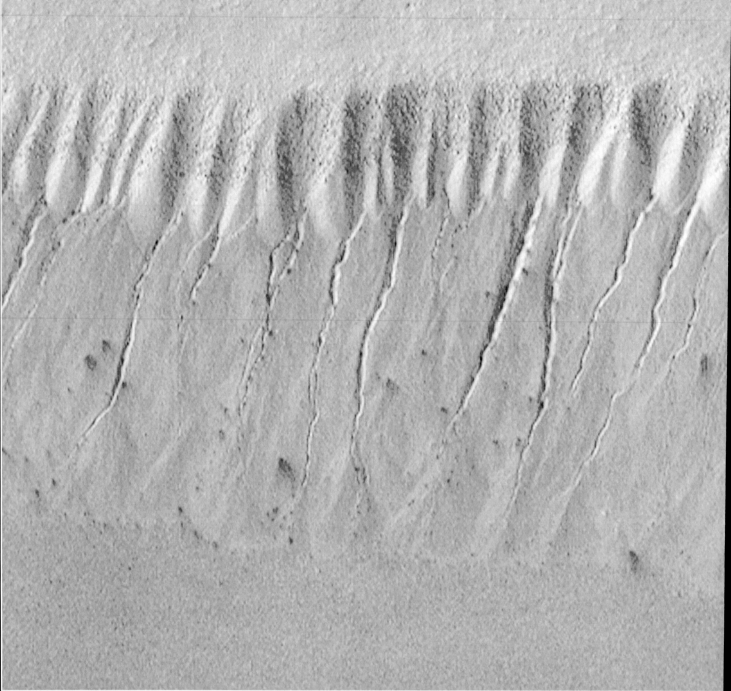

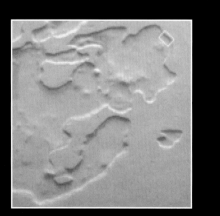

Detailed view from the Mars Global Surveyor (above) includes gullies with sharp, deep, V-shaped channels far too small to be seen in a Viking 2 photo (left; box indicates the MGS image location). The gullies occur in an unlikely place: the Martian equivalent of Antarctica, where temperatures remain well below freezing all year round. Ice here may possibly protect liquid water from evaporation until enough pressure builds for it to be released catastrophically down a slope.

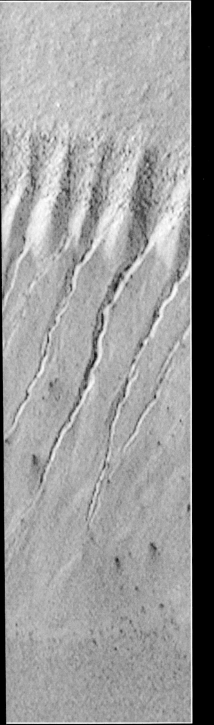

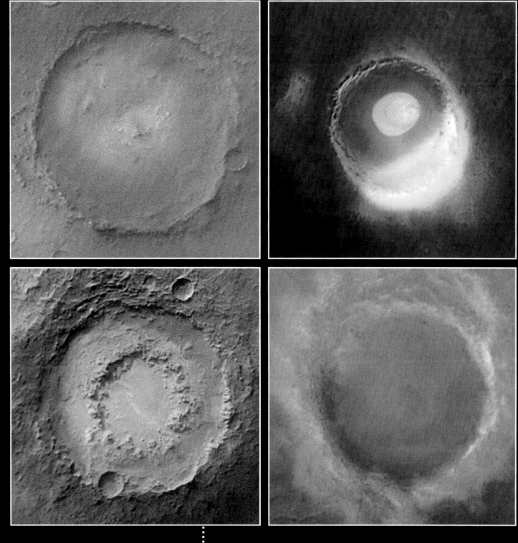

Different craters chronicle the passing seasons on Mars. In the northern hemisphere spring, frost accumulated during the six-month winter begins to retreat (top left), though one unnamed crater (top right) retains a patch that will persist through summer. At the same time, it's autumn in the southern hemisphere, encouraging frost to build in Lowell Crater (above left) and Barnard Crater (above right).

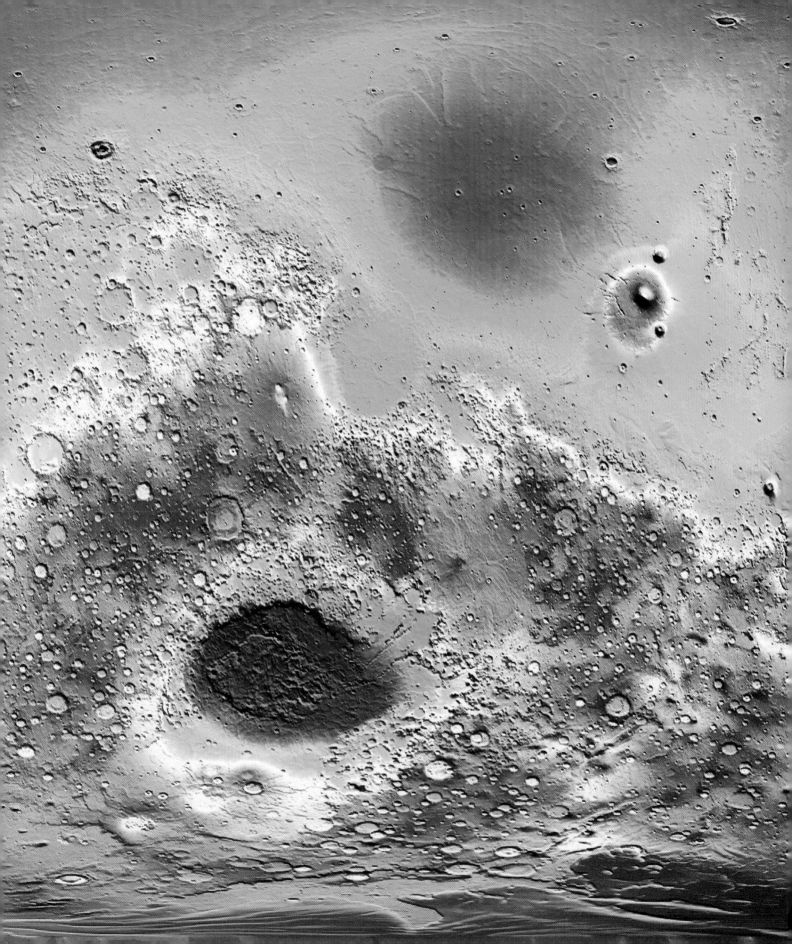

PIERCING EUROPA'S ICY SKIN

What Lies Beneath?

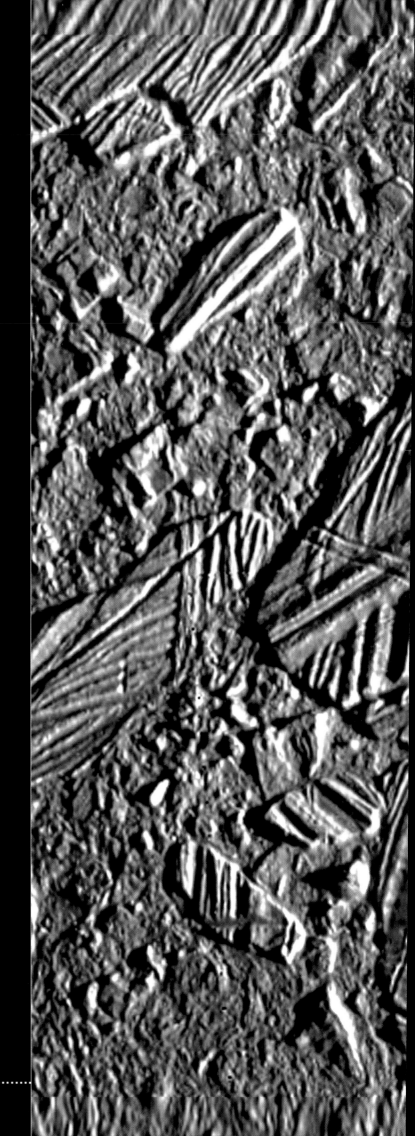

Despite a surface temperature of minus 260 degrees Fahrenheit, Jupiter's ice-covered moon Europa is one of the hottest spots in the solar system—from a scientific perspective. That's because recent data from NASA's Galileo probe strongly suggest that this moon's icy crust could conceal the solar system's biggest ocean, a vast subsurface layer of liquid water warmed both by radioactive decay in Europa's rocky core and by the crushing grip of Jupiter's relentless gravity.

Water and a source of energy are two of the three ingredients necessary for life on Earth and, by extension, anywhere else in the solar system. The third is organic material, and Europa probably has that in abundance from past bombardment by organic-rich comets and meteoroids. This does not mean life actually exists on Europa, of course. But it's more

Rafts of ice hide what could be the solar system's largest ocean, on Jupiter's moon Europa. The Galileo spacecraft confirmed Europa's icy crust. White areas have been dusted by ice particles ejected by a huge impact 600 miles away. Europa's elliptical orbit around Jupiter produces geologic stresses and heat that may create a slushy or even liquid interior layer.

To launch in 2008, Europa Orbiter will begin beaming radio waves through the ice of Jupiter's fourth largest moon by 2011. It should be able to detect icewater—perhaps as little as half a mile below the surface. Instruments will chart surface and interior processes. Later unmanned "hydrobot" missions may penetrate the ice and explore this moon's possible interior sea.

....

shape of the planet changes," Johnson says. "If you can measure a three-and-a-half-day change in the gravity field, you can tell how much mass is being moved into and away from the center." A laser altimeter could be used to measure the spacecraft's altitude as it circles the moon. "If the ocean on Europa is frozen solid," Johnson adds, "then as Europa goes around Jupiter the tide will basically rise and fall just a meter [3 feet] or so. If it's got a liquid layer there, the tide every three and a half days as you go around Jupiter will rise and fall by more like 30 meters [100 feet]."

The second goal of the Europa Orbiter is to thoroughly analyze the structure of the moon's interior in three dimensions and to characterize the overlying ice layer or layers. "How deep are the ice layers, how much does the depth of the ice vary, is there layering in the ice and does that tell us something about how it was laid down or stripped off?" asks Rob Staehle, deputy project manager of the Europa Orbiter team at the Jet Propulsion Laboratory. Ice in a particular spot,

he adds, could vary f
miles thick. "There a
could postulate that
not going to know w
unless you can...ima

Europa Orbiter li
wavelength radar cap
to a depth of many m
measures the moon'
presence of an ocean
should determine jus

A third objective
occurred since Galile
potential landing site
Europa enthusiasts h
compact submersibl
overlying ice and exp
But "that's a really ta
will work only if the

"Failing that, I th
to look for more info
places where, assum
below, it has come u
"Maybe it's through
welling up. It might
oroid impacts. There
splashed up on the s
nism, last month or
sand years ago, all yo
down a couple of me
ice, stick it in someth
have a great sample

Whether that sa
life is an open questi
downplay. "If you re
now," he says, "every
sentence has the wo
cool thing. But I'm a
We did that with Vik
according to most of
find life. And the pre
a failure. The missio

....

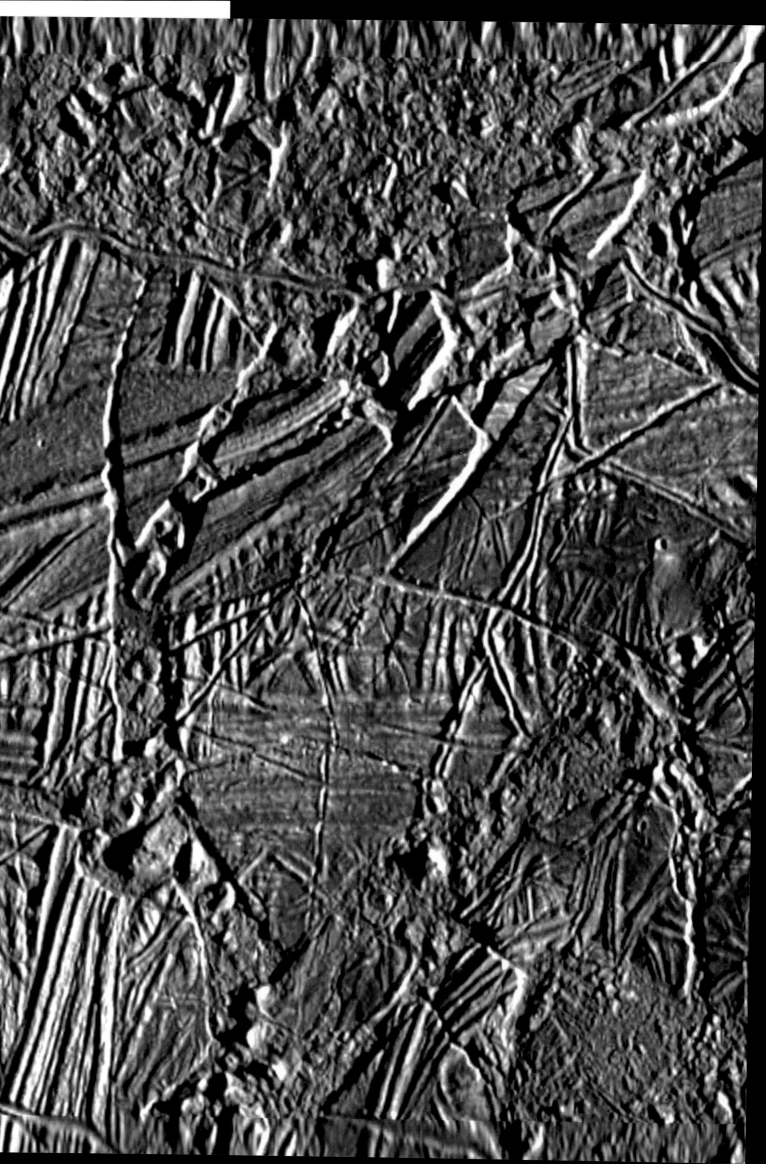

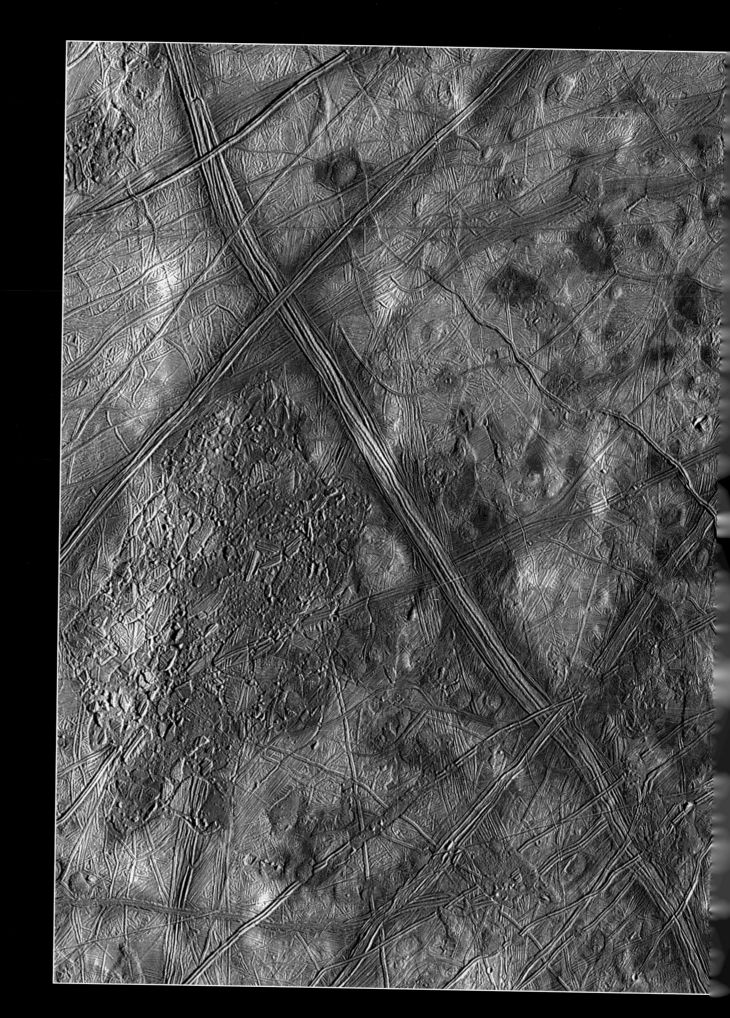

than enough to elevate this moon to a high-priority target—second only to Mars—in NASA's 21st-century planetary exploration program.

In one of the most ambitious and technically difficult robotic missions ever attempted, NASA plans to launch a nuclear-powered spacecraft in 2008 that will focus on proving, once and for all, whether a liquid ocean actually exists on Europa. If this orbiter finds such an ocean and if the overlying ice is not too thick, an attempt may be made in the next decade to launch a robotic lander that could make surface or even subsurface searches for signs of life. But it won't be easy.

"The Europa Orbiter is really pushing us," says project scientist Torrence Johnson. "If you want to get something down on the surface, you're talking about something that, in my personal estimation, is at least an order of magnitude more difficult than putting something on the surface of Mars. And we've found out THAT ain't all that easy."

Discovered in 1610 by Galileo, Europa is one of Jupiter's four largest moons, visible from Earth through small telescopes and even binoculars. Its diameter of 1,946 miles makes it only slightly smaller than our Moon (diameter: 2,156 miles); it orbits Jupiter some 416,900 miles from the planet's center, taking only 3.5 days to complete one circuit. Volcanic Io is closer, at 262,000 miles, thus it's subjected to even greater gravitational stress. Ganymede and Callisto orbit Jupiter at 663,400 and 1.2 million miles, respectively.

If these moons had perfectly circular orbits, the geologic stresses induced by Jupiter's gravity would be minimal. But because they're somewhat elliptical, the planet's gravitational grip varies as the moons wheel about, causing extreme tidal flexing. On Io, for example, this effect is so pronounced that the moon's surface rises and falls by up to 100 yards per day, says Johnson, continually generating the heat that drives its constant volcanism. Farther out on Europa, the effect is less pronounced. But it still generates an enormous amount of internal friction.

Decades ago, astronomers first detected the spectroscopic fingerprints of water ice on Europa's surface

by analyzing reflec
from NASA's Voya
a strange terrain w
remarkable absen
still-active resurfa
water ice and the s
with Jupiter's grav
whether Europa's
even liquid layer. T
One suggested tha
the other held that
to a frozen mudba

Then came Ga
with state-of-the-a
around Jupiter in
survey. During rep
analysis of how Eu
trajectory enabled
of the moon's inte

Now, says Joh
from whether or n
"the pizza-crust d
thick-crusters. You
young surface, we
don't have a froze
we've got 100 kilo
water on top of th

The thickness
future exploration
be able to answer.
it in 2008. By mid
Jupiter, using the
to "pump down" t
around Europa in
mission—to conf
ocean—must be
That's because the
in only one month
dosage that Galile
orbit around Jupit

Its mission ca
measuring how th

Intriguingly patterned with domes and ridges (opposite), Europa's ice rafts drift much like Earth's tectonic plates; reddish ridges mark their boundaries. Yet this moon is smoother than it looks; few features are more than a few hundred yards high, and most (below) are barely tall enough to cast shadows. One astronomer comments that they seem to be "painted on with a felt marker." But Europa's ocean—if it has one—may be 30 miles or more deep. Floating just above a Jovian ring (left), Europa makes a tempting target for NASA: It could harbor water, heat, and material from organic-rich comets and meteoroids— the three necessary ingredients for life as we know it.

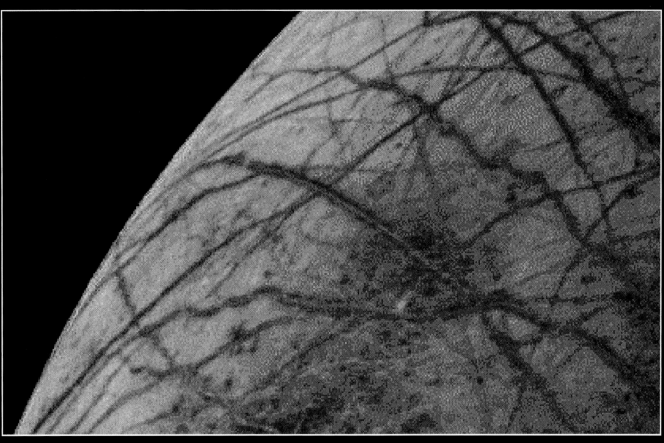

DESTINATION SATURN

Cassini and Huygens: A Team Effort

"**I**magine a world that's smaller than Mars and bigger than... Mercury, where the air is four times denser at its surface than the air in this room, and the surface pressure is about the same as you'd experience at the bottom of a neighborhood swimming pool."

It was September 1997, six weeks before launch of the most sophisticated interplanetary spacecraft ever built, and Jonathan Lunine, a planetary scientist at the University of Arizona, held a room full of reporters enthralled describing one of that mission's targets. "On that world," he continued, "the distant sun is never seen, and at high noon things are no brighter than a partly moonlit night on the Earth. Because of its great distance, the cold is so enormous that water is always frozen out of the atmosphere. Nitrogen is nearly so, but not quite. And the simplest

Nearing touchdown, the Huygens probe—built by the European Space Agency—descends to Saturn's moon Titan as NASA's Cassini mother ship streaks past at upper right; the ringed planet looms in the distance. For clarity, the artist has pared back Titan's dense hydrocarbon haze. Huygens should land in January 2005, making more than 1,000 images during its two-and-a-half-hour descent.

organic molecule, methane, is there to take the place of water as a cloud-former, possibly a rainmaker, and maybe even the stuff of lakes or seas of hydrocarbons."

Hundreds of miles above the surface of this world, methane is broken down by cosmic rays and the distant Sun's faint light, creating more complex organic compounds that rain down. Volcanism and meteoroid impacts shape the surface and provide the energy needed to create even more organic molecules, said Lunine, "in a planet-wide tapestry that is an organic chemist's dream."

The alien world he describes is not a planet but a moon: Titan, the largest satellite of Saturn and the target of a small probe hitchhiking to the ringed planet on NASA's nuclear-powered Cassini orbiter. While life almost certainly does not exist on Titan today, conditions there are thought to be similar to those on Earth shortly after the birth of the solar system, making Titan a high-priority target both for NASA and for the European Space Agency (ESA).

"The organic chemical cycles that go on may constitute a chemical laboratory for replaying some of the steps that led to life on Earth," Lunine said. "Titan is in some ways the closest analogue we have to the Earth's environment before life began, and this makes Titan very important."

Cassini and ESA's Huygens probe were launched atop an Air Force Titan 4B rocket in October 1997. Costing $3.4 billion through the end of its primary mission in 2008, Cassini is one of the most expensive interplanetary spacecraft ever built, second only to NASA's Viking Mars project in the 1970s. (Viking's official cost—$2.73 billion—was figured in 1973 dollars; that's nearly $4 billion in today's dollars.)

In an era of faster, better, and cheaper spacecraft, where quick development, modest mission objectives, and tight budgets are the rule, Cassini stands out as a glaring exception. Yet it is expected to deliver a tremendous bang for all those bucks. For starters, its primary target is 750 million miles from Earth at its closest, about 21 times farther than Mars is at its nearest approach. Just getting there takes not six months but

seven years, and involves a three-planet gravitational bank shot. Once there, Cassini will spend at least another four years studying the Saturnian system, accomplishing in one mission what would take more modest spacecraft several flights to match.

It will examine the planet's atmosphere, magnetosphere, and rings, as well as its many icy moons, said Wesley Huntress, NASA's director of space science at the time of Cassini's launch. "The mission represents a rare opportunity to gain significant insights into major scientific questions about the creation of the solar system, pre-life conditions on early Earth, and just a host of questions about Saturn."

If all goes well, Cassini will brake into orbit around Saturn on July 1, 2004. Huygens will descend about six months later, spinning at seven rotations per minute for stability. It will hit the moon's upper atmosphere at an altitude of about 750 miles and a velocity of 13,725 mph. In just three minutes, the probe will slow to less than 900 mph, experiencing heat-shield temperatures up to 21,600 degrees Fahrenheit and a braking force that's 16 times the pull of Earth's gravity.

A small pilot chute will deploy, pulling out Huygens's 27-foot-wide main chute at an altitude of about 100 miles. For the next 15 minutes, the spacecraft will slowly descend while its instruments make initial observations of Titan's atmosphere. Then, at an altitude of about 70 miles, the main chute will be cut away and a 10-foot-wide stabilizer chute will open. Huygens should hit the surface at 15 mph or so, two and a half hours after the descent begins. Data collected throughout the descent and landing will be beamed up to Cassini for relay to Earth.

The probe is expected to make more than a thousand images of Titan's surface and clouds, also to make spectroscopic measurements and monitor the penetration of sunlight. A different instrument will measure wind velocity, while yet another will analyze atmospheric gases. An onboard weather station will not only measure temperature, pressure, and electrical activity but also look for lightning. Finally, a surface science package equipped with nine different sensors will try to determine the physical nature of whatever surface the probe lands on, liquid or solid, and should provide a point of reference that will help scientists interpret the radar imagery that will be carried out by Cassini itself.

"We hope, we have good confidence, the probe will survive landing," ESA project scientist Jean-Pierre Lebreton said before launch. "The landing speed is very low and...we have capability to do measurements for half an hour on the surface."

Once the Huygens mission is complete, the science team will settle down for the most comprehensive study of Saturn and its moons ever attempted. Cassini, by utilizing Titan's gravity to alter its trajectory, will make 70 ever-changing orbits, accomplishing repeated close flybys of half a dozen moons.

Earl Maize, Cassini spacecraft operations manager, says the spacecraft will make Saturn "one of the most intensely observed planets in the solar system. Our instruments run from the infrared to the deep ultraviolet; we have almost a full spectrum of fields and particles instruments, dust and atmospheric constituent [analyzers]. It really will run the range. We intend to spend four years and, hopefully, if all goes well, go on for an extended mission."

Cassini's 18 instruments and 27 sensors include cameras to make close-up and wide-angle photographs of Saturn, its larger moons, and its glorious ring system. A visible and infrared mapping spectrometer will chart the distribution of specific minerals and chemicals in the atmospheres of both Saturn and Titan, as well as in the planet's rings and on the surfaces of Titan and other moons. It also will search for lightning and active volcanoes on Titan. A composite infrared spectrometer will measure infrared emissions from each target—the rings, the surfaces, and the atmospheres—to produce vertical temperature profiles and to help determine atmospheric composition. A third spectrograph will use ultraviolet imaging to determine how much hydrogen and deuterium are present. A remote-sensing microwave package includes a powerful radar system capable of seeing through Titan's cloud cover and imaging features as small as 1,150 feet across. Six other instruments will probe Saturn's magnetic field and assess how it interacts with the Sun's solar wind, the energetic plasma associated with Saturn's magnetic field, and the particles trapped within it.

That's a lot of scientific horsepower to invest in a single spacecraft. But if any planet deserves red-carpet treatment, surely it is Saturn. While all four of the gas giants—Jupiter, Saturn, Uranus, and Neptune—feature complex ring systems, Saturn's is by far the most spectacular. It's also the only one easily visible from Earth. A backyard telescope reveals two bright rings separated by a dark gap called the Cassini division. Large observatory instruments reveal a half-dozen clearly defined ring regions and four gaps. But their true nature was not revealed until NASA's Voyager probes flew past in 1980 and 1981. To the delight and amazement of astronomers, the Voyager cameras revealed more than a thousand individual rings whirling about the planet in a complex, ever changing dance orchestrated by the gravity of Saturn and a handful of small moons.

Composed of icy particles that range in size from tiny motes to boulders 16 feet or so across, Saturn's most clearly defined rings extend more than 60,000 miles from their innermost boundary to their outer reaches. Despite their large radial size, the rings are tissue-paper thin, relatively speaking, just 300 feet or so thick at most.

The planet itself, second only to Jupiter in volume, is big enough to hold 750 Earths. Its density, however, is less than that of water and its overall mass is just 95 times that of Earth. Long-lasting hurricane-like storms and a 1,100-mph equatorial jet stream roil its upper atmosphere. Saturn emits 87 percent more energy than it absorbs from the Sun, possibly due to the friction of liquid helium that rains down through layers of hydrogen that compose its deep interior. All in all, it makes a beautiful, if inhospitable, place to visit. Should space travel ever become routine, the major hotel chains surely will flock to Saturn. Just imagine the views from an orbiting hotel. Then imagine the rates for "ringside" rooms. Start saving now.

Fully assembled and ready to go, Cassini (above) undergoes numerous checks in 1997 in Pasadena, California. In October of that year it successfully lifted off from Florida (opposite). After a 1.1-billion-mile journey, the spacecraft will reach Saturn by July 2004. It will spend four years in Saturn's neighborhood (above right), using 18 instruments and 27 sensors to study the planet's atmosphere, magnetosphere, and rings, as well as its many moons. The Huygens probe, meanwhile, will descend to Titan's cloud-shrouded surface (below), relying on battery power to radio Cassini the atmospheric and surface data it gathers; Cassini will relay that information and its own to Earth. Astronomers are interested in Titan because conditions there may be similar to those on Earth billions of years ago.

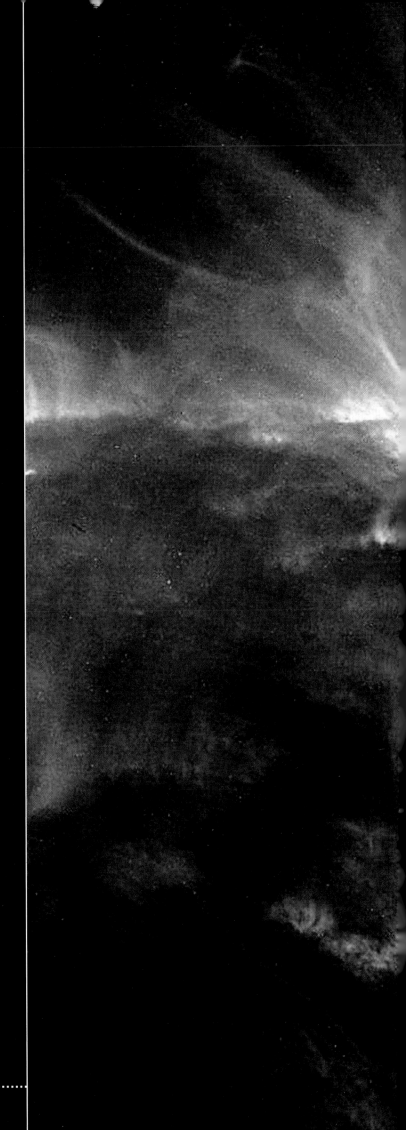

Laying Bare the Sun

I n 1989, one of the most powerful explosions ever recorded in the Sun's outer atmosphere sent a vast cloud of electrically charged particles hurtling toward Earth like birdshot from a cosmic shotgun blast. Two days later, that blast slammed into our planet's protective magnetosphere, triggering spectacular auroral displays as showers of electrons and protons spiralled down magnetic field lines and crashed into the atmosphere above Earth's poles. The resulting geomagnetic storm did much more than put on a dramatic light show, however. It also caused equipment failures in a Canadian power grid that left six million people in the dark for nine hours.

No one knows exactly what triggers the magnetic short circuits that apparently cause the Sun's million-degree corona to suddenly generate such cataclysmic flares and even

Solar fireworks erupt in March 2001, as our home star reaches a cyclical peak. Oscillating coronal loops—a rare phenomenon—occur when a filament eruption causes mayhem in the corona. TRACE, one of a new generation of instruments studying the Sun, caught this early phase of the eruption. Such cataclysms not only affect Earth short-term but also may contribute to long-range phenomena like global warming.

more powerful coronal mass ejections (CMEs)—mind-boggling explosions that periodically spew billions of tons of ionized gas into space. But a new generation of satellites is working around the clock to collect data needed to answer the Sun's most puzzling riddles and to provide an early warning that would minimize the Earthly effects of flares and CMEs.

"It's one thing to say a line of thunderstorms is coming," says George Withbroe, director of NASA's Sun-Earth Connection division. "It's another thing to say, 'Oh by the way, that line of thunderstorms includes tornadoes.' We want to improve the reliability of the predictions."

For good reason. A single solar flare can unleash the energy equivalent of more than a billion thermo-nuclear bombs, accelerating charged particles to near the speed of light and increasing the temperature of an Earth-size region of the Sun by tens of millions of degrees. Waxing and waning with the Sun's 11-year sunspot cycle, flares may or may not be accompanied by CMEs. Along with disrupting power grids, both flares and CMEs can scramble communications, cause satellite malfunctions, and pose a radiation hazard to orbiting astronauts.

Even small variations in the Sun's overall energy output also can have a profound long-term effect on Earth. In the mid-17th century, for example, flares and CME activity virtually ceased for more than 50 years. The Sun's total output dropped by only a quarter of one percent, but that was enough to produce the Little Ice Age, marked by especially harsh winters and a brief expansion of glaciers in the Swiss Alps.

Understanding the Sun's behavior will help us predict possible consequences to our own climate, which could go far beyond the onset of another ice age. "If there is long-term variability in the Sun that can either increase or decrease global warming, that's important to understand," Withbroe says. "There are folks who study these issues who think maybe up to 30 percent of global warming in the past century is due to the Sun getting brighter. Most of the warming appears to be caused by human activity, but if the Sun has contributed

Size of the Sun

Some 62 million miles wide, a gigantic hydrogen cloud surrounds comet Hale-Bopp, greatly exceeding the comet's visible tail (inset). This image comes from SWAN—Solar Wind Anisotropies, an instrument aboard SOHO that looks for hydrogen atoms coming from interstellar space. Astronomers estimate that, when this image was made, Hale-Bopp was losing about 55 million tons of water vapor a day.

30 percent, that's important to know if you're trying to use the past in order to predict the future."

Enter SOHO—the Solar and Heliospheric Observatory—a European Space Agency satellite launched by NASA in December 1995 that has quietly revolutionized knowledge about the Sun's internal structure and how it interacts with the seething solar atmosphere. Equipped with a battery of light-splitting spectrometers, telescopes, and solar-wind monitors, SOHO orbits a point about one million miles from Earth, a gravitational eddy where it can remain with minimal effort. After five years of operation, SOHO still has enough fuel for another ten years of observations.

SOHO was designed to study the entire Sun, from its hidden core—where nuclear fusion reigns supreme—to its convective outer layers—where flares and CMEs are born and where the solar wind blows out into space. To probe the Sun's interior, SOHO monitors ultralow-frequency sound waves that cause vast stretches of the visible solar surface to gently move up and down a few hundred yards every few hours. These sound waves, generated by surface convection, propagate throughout the Sun's interior and return to the surface, causing the Sun to vibrate like a gong.

How fast a given sound wave moves through the interior depends on its frequency as well as the temperature and composition of the material it passes through. By studying the precise nature of surface vibrations, helioseismologists can infer details about the Sun's interior structure and gain insights about what drives the magnetic dynamo that presumably heats the corona and powers flares and CMEs.

Like Earth, the Sun is layered; its thermonuclear core extends from the center to about 25 percent of the Sun's radius. The so-called radiative zone, where energy is transported toward the surface by radiation, extends to 71 percent of the Sun's radius. From there to the surface, energy is carried by turbulent convection. Data from SOHO indicate the radiative zone rotates rigidly, as if it were solid. The convective zone, however, rotates differentially; areas near the equator come full circle every 25 days or so, while regions at higher latitudes spin more slowly.

"There's a lot of shear going on right there at the interface," Withbroe says. "That's a good place for magnetic fields in motion to wind up. So ultimately, the studies of the interior are going to give us clues as to how the whole dynamo works." Magnetic energy generated by that dynamo is believed responsible for heating the corona to its extreme temperatures. In a 60-mile-thick transitional zone between the Sun's lower atmosphere and its corona, the temperature jumps from 10,000 kelvins to 1 million.

NASA's Transition Region and Coronal Explorer satellite (TRACE), launched in April 1998, studies the Sun's outer atmosphere with a high-resolution imaging system designed to complement SOHO's data. In its first year of operation, TRACE took more than 1.5 million images of the Sun, providing new information on the mystery of coronal heating. Building on such successes, the United States, France, Germany, and the United Kingdom plan to launch STEREO—the Solar Terrestrial Relations Observatory—in 2004. It will consist of two identical spacecraft designed to study coronal mass ejections in three dimensions. NASA also had hoped to send another spacecraft, the Solar Probe, to within 1.2 million miles of the Sun, actually sailing through the outer reaches of the solar atmosphere. But that $600-million project was cancelled to free up money for other programs. Now NASA's next major Sun spacecraft will be the Solar Dynamics Observatory (SDO), which will concentrate on high-resolution helioseismology from Earth orbit.

"To really do helioseismology, you need a tremendous amount of data," says Withbroe. "We'll put SDO in geosynchronous orbit [22,300 miles above Earth's equator] and have a dedicated antenna on the ground and take just huge quantities of data down." Scientists, he adds, will be able to "look below a sunspot, see its roots. You can see the roots of an active region, follow it as it crosses the surface of the Sun. And we also have the capability to observe sound waves on the back side of the Sun, and see what's going on on the other side."

NASA hopes to launch SDO at the end of 2006. It will work in concert with Japan's Solar B spacecraft, scheduled to carry out in 2005 high-resolution observations of the corona in visible, ultraviolet, and x-ray wavelengths. The two satellites represent the next generation of solar research, and they may eventually answer another enduring scientific puzzle: What drives the 11-year solar cycle?

SDO will be the flagship mission in NASA's "Living with a Star" program, which Withbroe devised. In addition, NASA plans to launch a small fleet of modest Solar Sentinel spacecraft in time for the next solar maximum, in 2011, to observe the far side of the Sun, to characterize the evolution of CMEs and solar flares, and to improve the general forecasting of solar storms.

Says Withbroe, "There are two places in the solar system we haven't been to—the Sun and Pluto. And the Sun is really unique. It's the only star we're going to be able to go visit for a very long period of time—unless we get some new physics."

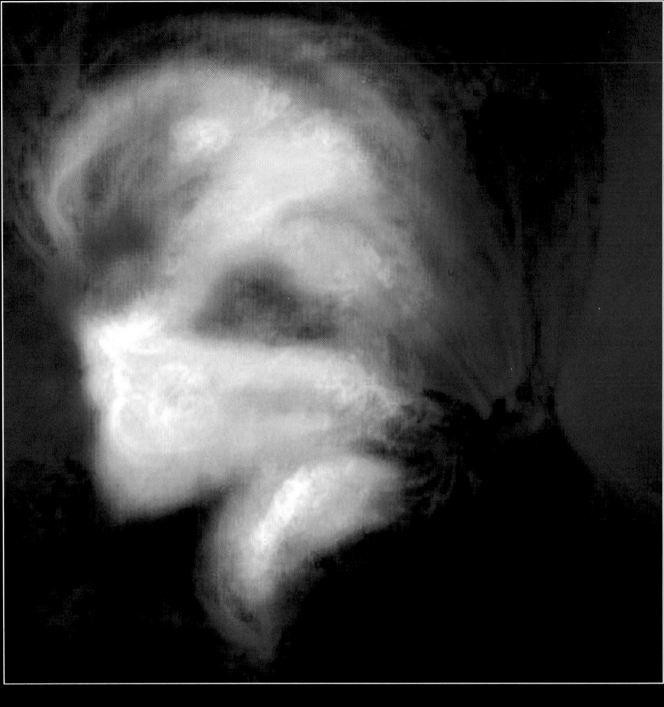

Celestial "moss"—a newly discovered feature named for its spongy appearance—takes root on the Sun just a couple of thousand miles above the top of the chromosphere, or solar surface, and below the much hotter and more expansive material of the corona. Data from TRACE and a Japanese satellite launched in 1991 called Yohkoh—its name means "sunbeam"— were combined to make this image. Skylab, now long gone, also contributed dramatic observations of our Sun. During 1973 and 1974, three teams of astronauts aboard the U.S. space station made solar observations—specifically x-ray imaging—a significant part of their duties. They were first to detect temperatures of over 100 million degrees in flares, and also discovered that the corona was much more structured than previously expected.

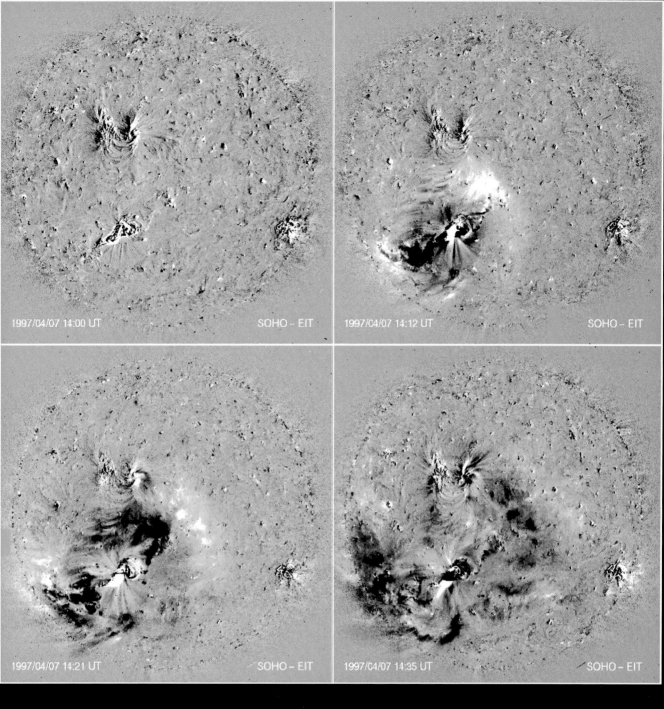

1997/04/07 14:00 UT SOHO – EIT

1997/04/07 14:12 UT SOHO – EIT

1997/04/07 14:21 UT SOHO – EIT

1997/04/07 14:35 UT SOHO – EIT

Traveling at 185 miles per second, a disturbance known as a Moreton wave—triggered by a coronal mass ejection—expands across the Sun's surface in sequential photos (above) that span 35 minutes. They were taken by EIT—the Extreme Ultraviolet Imaging Telescope—aboard SOHO in May 1997. Extreme solar activity blasts Earth's face with a magnetic cloud of plasma 30 million miles wide (right), spawning storms and electrical interference on the planet. White lines define the solar wind, violet the bow shock wave; blue represents Earth's ever-protective magnetosphere, bent but not broken by the assault.

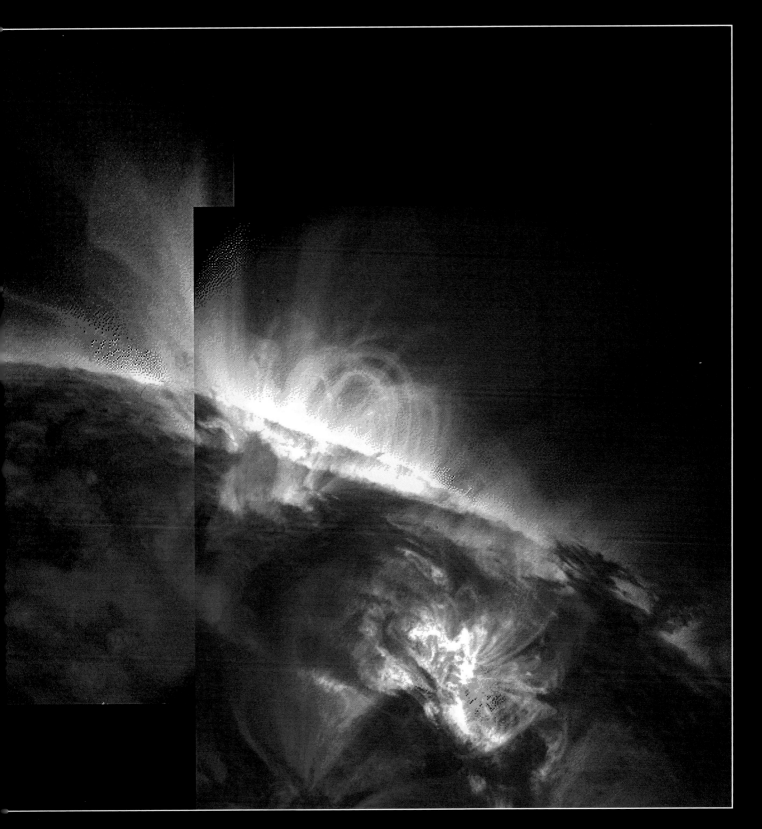

High-arching loops connect two active regions on either side of the Sun's equator. Relatively cool plasma shows as red, warmer as pink. NASA's future solar studies hopefully will include the Solar Dynamics Observatory, which will specialize in high-resolution helioseismology from a near-Earth orbit. In 2004

NASA also plans to launch two spacecraft jointly with several European countries. Called STEREO—for Solar Terrestrial Relations Observatory—it will create 3-D images of coronal mass ejections. Five years after that, a fleet of sentinel spacecraft will venture out from Earth to observe the far side of the Sun.

Homing in on Tiny Targets

Missions to Comets and Asteroids

I n the cold predawn hours on January 15, 2006, a helicopter whisks a recovery team across Utah's salt flats, zeroing in on a long-awaited UHF radio beacon. Suddenly the chopper's searchlight spots a collapsed parachute. On the frigid ground nearby lies the goal: A 32-inch-wide saucer-shaped capsule with a charred heat shield and a priceless cargo. Inside, locked within tiny beads of smoke-like aerogel, are particles of comet dust, pristine remnants of the primordial cloud of stellar debris that coalesced to form our solar system more than four and a half billion years ago.

The capsule's sample container is carefully removed and flown to the Johnson Space Center in Houston, where eager scientists begin an exhaustive battery of tests. No one knows what they will find. But they expect the tiny grains of ancient star stuff will open a new

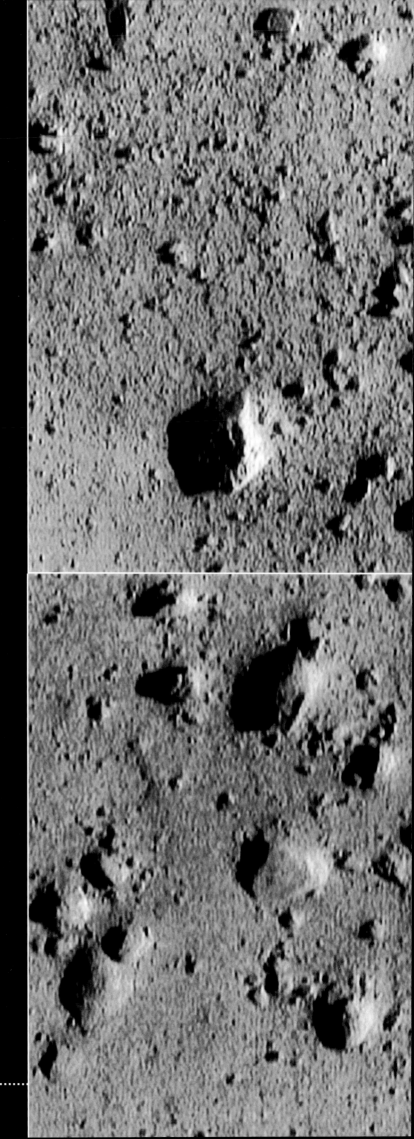

Closer and closer, NEAR Shoemaker approaches the asteroid Eros, touching down on February 12, 2001. The flight's final four images range in altitude from 3,773 feet (upper left) to 394 feet (lower right); the last encompasses an area just 20 feet across and bears marks of interference caused by the landing. Astronomers believe that comets and asteroids may hold secrets of the universe previously unsuspected.

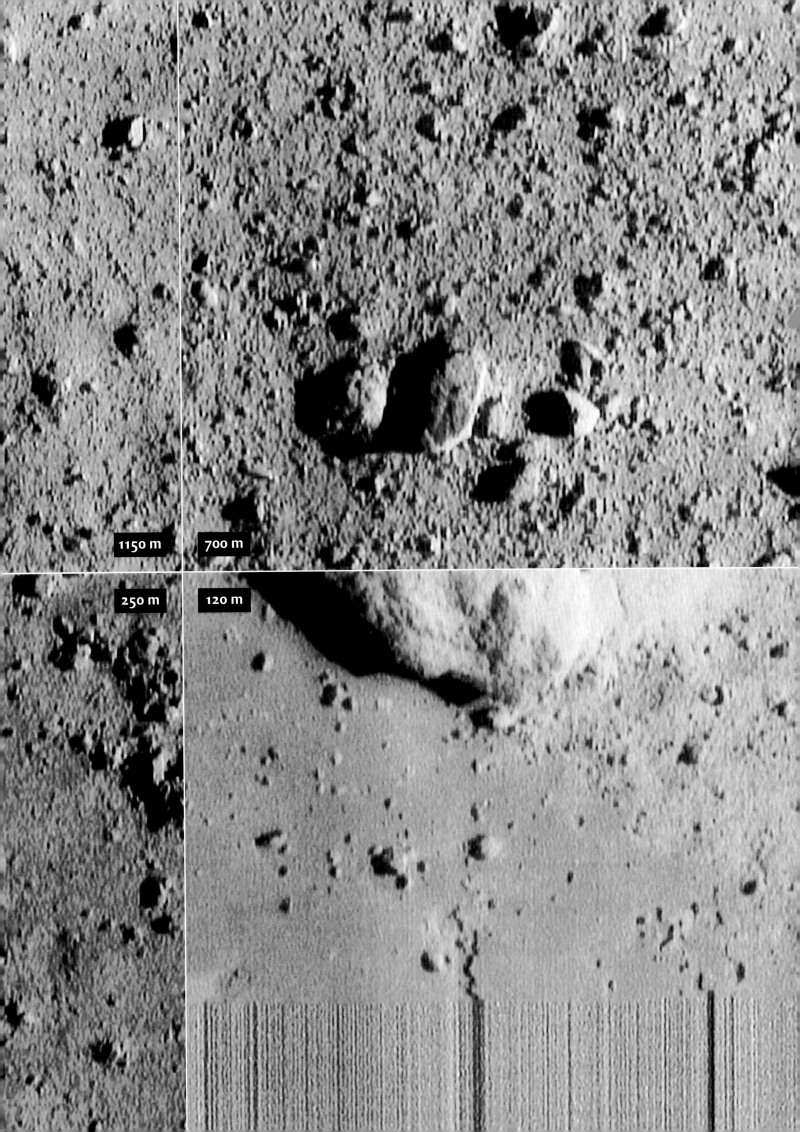

1150 m

700 m

250 m

120 m

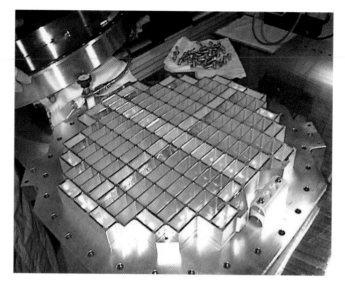

window on the birth of the solar system and, quite possibly, the evolution of life on Earth.

"Comets are the stuff that life is made of," says Carl Pilcher, now NASA director of the Solar System Exploration division. "The organics we find in meteorites, which we think are similar to the organics we would find in comets, contain lots of amino acids, and amino acids are the building blocks of proteins. So comets probably delivered lots of amino acids to Earth—as well as enough water to fill Earth's oceans. The fact that life on Earth formed very quickly after the Earth formed itself is probably due to all this material that was brought into Earth by comets. Comets probably delivered this same material to Venus and Mars as well, so comets are a link between our study of the formation of the solar system and our study of the biological potential of Mars."

While missions to Mars, Jupiter, Saturn, and their moons are the most publicly visible unmanned space projects, NASA, the European Space Agency, and Japan's Institute of Space and Astronautical Science are kicking off the 21st century with an unprecedented survey of comets, asteroids, and the solar wind, hoping to fill in blanks about the solar system's early history.

By 2008, researchers also expect to complete a ten-year project to locate comets and asteroids whose orbits carry them close enough to Earth to pose a potential threat. Don Yeomans, an asteroid expert at the Jet Propulsion Laboratory, explains the goal: "to find, within ten years, 90 percent of those near-Earth objects larger than one kilometer [0.6 mile]. That's sort of a dividing line between a local and a global disaster." Although tons of extraterrestrial material

Stardust, the first spacecraft designed to explore a comet and return to Earth with cometary material, takes off in February 1999 (above left). Aboard are cells of aerogel (above)—a silicon-based solid, 99.8 percent air and a thousand times less dense than glass—that can trap and cushion tiny particles. By early 2004, Stardust should approach the tail of comet Wild-2 (opposite) and begin gathering samples of its debris. Later analysis on Earth could yield insights into the evolution of the Sun and planets— and perhaps even the origin of life.

bombard Earth every day, most is dust or pebble-size, which burns up in battling through our atmosphere. The smallest objects to actually reach Earth's surface strike at roughly 100-year intervals; those big enough to spark global consequences hit every 300,000 years or so. Comets and asteroids, Yeomans adds, "are not just the flotsam and jetsam of the solar system. They really are, next to the Sun itself, probably the most important objects out there in terms of their power over life."

On its way to Jupiter, NASA's Galileo probe kicked off what has become a golden age for comet and asteroid research. It beamed back the first high-resolution photos of an asteroid as it flew past Gaspra in October 1991 and photographed asteroid Ida two years later, discovering that Ida sported its own tiny moon, since named Dactyl. The Cassini spacecraft also got in on the asteroid act, sending back photographs of Masursky in January 2000 as it crossed the asteroid belt on the way to Jupiter and Saturn. NEAR—the Near Earth Asteroid Rendezvous spacecraft—flew past Mathilde in 1997

and spent two years in orbit around Eros, before ending its mission with a dramatic soft landing on the asteroid's cratered surface, in February 2001.

Don Burnett, principal investigator for the Genesis mission scheduled to launch in July 2001, hopes to collect pristine samples of the solar wind, which is believed to be representative of the original nebula that evolved into the Sun and planets. Genesis is slated to return those samples to Earth in mid-2004.

NASA's Comet Nucleus Tour, or CONTOUR, will carry out photo reconnaissance of two comets at the height of solar heating, in June 2006. NASA also plans to sample the interior of a comet with a probe called Deep Impact, which launches in January 2004. After catching up with its quarry—Comet Tempel 1—by July 2005, it will fire a 770-pound projectile at the surface, excavating a crater the size of a football field. Onboard instruments then will study the chemical composition of the resulting cloud of debris.

Japan intends to launch a sample-return mission in July 2002 that will rendezvous with a 1,300-foot-wide asteroid known as 1989ML. Its probe will fire up to three bullets into 1989ML's crust and capture whatever debris results. An innovative solar-powered electric propulsion system will return samples of the asteroid to Earth in June 2006.

The European Space Agency plans the most ambitious mission, an eight-year voyage to Comet 46 P/Wirtanen starting in 2003. Its Rosetta spacecraft will fly in tandem with the comet for two years and dispatch a small lander to drill beneath its surface. The orbiter will carry four remote-sensing cameras and spectrometers, plus six instruments devoted to analyzing surface composition; the lander will have its own camera, an x-ray spectrometer, two gas analyzers, and a magnetometer, as well as subsurface probes.

"As we cannot bring the cometary material into our terrestrial laboratories, we will take our laboratories to the comet," ESA science director Roger Bonnet declared in 1993, when the project was first approved.

Now NASA's Stardust spacecraft, launched in February 1999, is on its way to do just what Bonnet said wasn't possible: Collect material from a comet's coma and bring it to Earth for detailed laboratory analysis. Using a gravity-assisted flyby of Earth to boost its velocity, Stardust should reach comet Wild-2 early in 2004. On its way there, it will also collect dust grains from interstellar space.

Buffeted by particles speeding along at 22,320 mph—six times faster than a rifle bullet—the armored spacecraft will pass within 63 to 93 miles of Wild-2's icy nucleus, snapping high-resolution photos and capturing particles of its dusty coma in a cushion of glass fiber aerogel. The particles will bore into the aerogel like tiny bullets, melting it and, in effect, encasing themselves in minuscule glass beads. It will take Stardust another two years to return and drop off its payload for a parachute descent to Utah.

Wild-2 is a particularly good target because it has kept mostly to the outer solar system, where it remains largely unaffected by solar heating and surface vaporization. Whatever material it yields should be representative of the nebula that gave birth to the solar system. Says Stardust senior scientist Martha Hanner, "The comets we see today have been essentially parked in cold storage for much of the four-and-a-half-billion-year history of our solar system. So they're truly frozen time capsules; that's why we're so interested in them."

FIRE AND ICE:
GOING BACK
TO MERCURY;
OFF TO PLUTO

Hellish Mercury and frozen Pluto frame our solar system like cosmic bookends. At one extreme, Mercury zips around the Sun every 88 days at an average distance of 36 million miles. Pluto glides serenely through space a staggering 3.7 billion miles out, taking 248 years—longer than the United States has been an independent nation—to complete a single highly elliptical orbit.

Moonless and cratered Mercury, visible to the unaided eye, possesses an oversize metal core that makes it nearly as dense as Earth. Pluto, visible only through large telescopes, is an icy, volatile-rich snowball slowly rotating in lockstep with a moon not much smaller than itself. Surface temperatures on Mercury vary from -300°F in the shade to a sizzling 800°F in direct sunlight, hot enough to melt lead. On Pluto, the average temperature hovers around 40 degrees above absolute zero, nearly -400°F.

From 2.6 billion miles, Hubble easily discerns Pluto and its moon Charon, which most ground-based telescopes blur together. It's a feat similar to distinguishing a baseball from 40 miles away, and it's enabled scientists to accurately measure the diameters of both— about 1,440 miles for Pluto and 790 miles for Charon.

So cold, in fact, that the planet's tenuous atmosphere periodically condenses and settles to the surface.

Despite their many differences, Mercury and Pluto share at least one common trait: Less is known of their structure and evolution than any other planets in the solar system. And yet both could hold the keys to understanding the early history of the swirling cloud of gas and dust that gave birth to the Sun and its retinue of planets, comets, and asteroids 4.6 billion years ago.

No spacefaring nation has ever sent a probe to Pluto, so far away that any currently envisioned spacecraft would need seven to ten years just to cross the vast interplanetary gulf. Even then, scientists would have to settle for a quick flyby, because no currently available rockets can carry enough fuel to brake into orbit around relatively light and slow-moving Pluto.

It is almost as difficult to reach fleet Mercury. A single spacecraft—NASA's Mariner 10—flew past the planet three times in 1974-75, but mapped just 45 percent of its surface, in large par because Mercury rotates so slowly—only once every 59 Earth days. The planet's other 55 percent remains a mystery. But NASA plans to send another spacecraft there in April 2004. Shielded against the blazing heat of the Sun, the $286-million MESSENGER probe will photograph most of Mercury's surface during two trajectory-warping flybys, in 2007 and 2008, before finally slipping into orbit in 2009 for a full year of up-close observations.

"Seeing the other side of Mercury is only one of the reasons we want to go back," says Sean Solomon, principal investigator of the MESSENGER project. "We've learned a lot more about how to make measurements around planets and, of course, we've never sent an orbiter to Mercury."

It has proven more difficult to mount even a flyby mission to Pluto, a target Rob Staehle, of the Jet Propulsion Laboratory, calls "the Mount Everest of space exploration." After intense lobbying by Staehle and others, NASA began studying proposals for a Fast Pluto Flyby mission in 1991. This idea evolved into the $500-million Pluto-Kuiper Express, which called for a more scientifically capable spacecraft. Launch was tar-

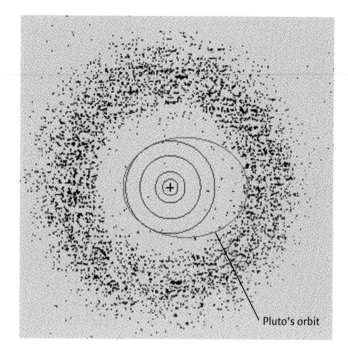

Pluto's orbit

Snugly cinched, the Kuiper belt—a flat ring of thinly scattered lumps of ice and other material—grazes the orbit of Pluto, which it may have spawned. NASA had planned a mission here, the Pluto-Kuiper Express, but put it on hold in 2000 due to budgetary problems.

geted for 2004, and the goals were to photographically map Pluto and its moon, Charon, to characterize their surface geology and composition, and to chart the constituents and behavior of Pluto's thin atmosphere.

But NASA issued a stop-work order on the Pluto project in 2000, citing rising costs, a relatively static budget, and the agency's commitment to high-priority targets like Mars and Europa. Disappointed scientists and engineers went back to their drawing boards, and today are hopeful that some sort of mission to the most remote planet will be launched before this decade is out. Says Staehle, "It's a fascinating place. There's a lot to be learned out there."

The same is true of the solar system's other extreme. MESSENGER will circle Mercury in a 12-hour elliptical orbit with a low point of 124 miles and a high point of more than 9,000 miles. Data will be collected as the probe passes close by the planet, then relayed back to Earth during the high-altitude portion of every other orbit. Orbital observations are expected to last a full Earth year. Among the questions MESSENGER will attempt to answer are why this planet is so dense, how its core is structured and how it powers Mercury's

Patchwork portrait of Mercury consists of 18 images taken by Mariner 10 in 1974. Its hostile surface—hot enough to melt lead—is, with Pluto's, the least studied in the solar system. In 2009 a new spacecraft, MESSENGER, hopefully will examine it in detail.

magnetic field, and what the chemical composition is of its surface and its ultrathin atmosphere.

"Mercury holds some secrets to how all the planets in the inner solar system got put together," says Solomon. "We know its density is exceptionally high for a planet of its size and we know, therefore, that it's made mostly out of iron, the most abundant heavy element. So the current thinking is it's probably mostly a core of metal surrounded by a thin shell, maybe 600 kilometers (370 miles) thick, of rocky silicates. The question is, how did Mercury end up mostly metal?"

One possibility is that an object larger than our Moon crashed into the planet in the remote past, stripping away much of the original crust and silicate mantle. Another possibility is that shortly after Mercury

formed, the Sun went through a period of increased energy output that vaporized the planet's outer layers. Or perhaps as the solar nebula coalesced to form planets, lighter materials fell inward due to friction with nebular gas near the Sun, leaving mostly metals in the region that gave birth to Mercury.

"What is nice about the ideas is they make some different predictions for what the rocky part of the planet should look like chemically," Solomon says. "So we can take chemical remote-sensing instruments and answer very fundamental questions about how Mercury got put together and, by inference, what were the important processes that affected all the inner planets."

Another major objective for MESSENGER is to study Mercury's magnetic field, initially detected by Mariner 10. Such fields are thought to be caused by the movement of a liquid outer core swirling around a solid inner core. But because small bodies cool relatively quickly, scientists were surprised to discover that Mercury had any magnetic field. MESSENGER will use a magnetometer to chart the structure and strength of the field. At the same time, its laser altimeter will collect data to precisely map the planet's topography and subtle variations in its rotation.

"Theory says the amplitude of that variation is going to be bigger if there's a fluid core than if the entire planet is solid," Solomon explains. "If there's a fluid outer core, it will have an amplitude about twice as big as if the core is solid."

Earth-based radar observations of Mercury indicate the possible presence of ice in polar craters, where the Sun's blazing heat never reaches. Such deposits could result from slow outgassing from the planet's interior or from past cometary impacts. Alternatively, they might be elemental sulfur, which can produce radar signatures similar to ice. MESSENGER should be able to resolve this riddle, using its neutron spectrometer to search for the hydrogen that would be present if the deposits are, in fact, ice.

"There are a host of very general processes that are at their extreme on Mercury," Solomon adds. "That's why we're going."

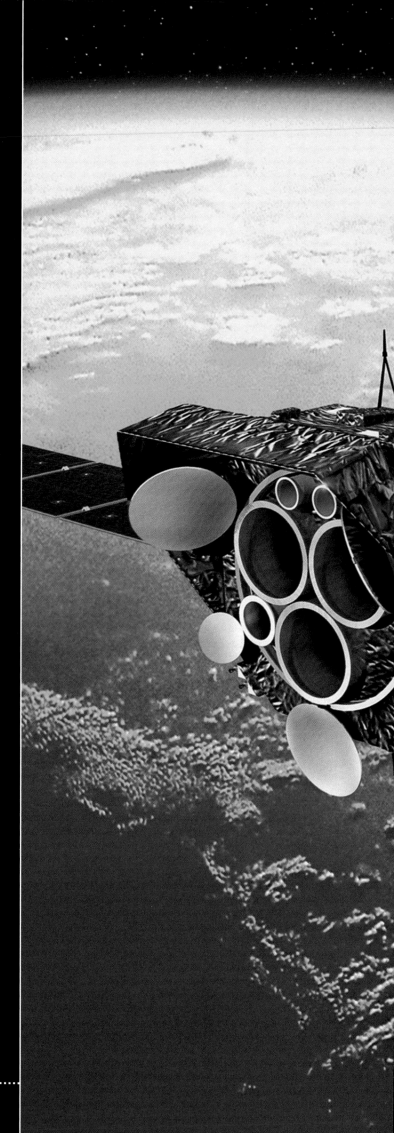

SPACE TELESCOPES COME OF AGE

Seeking the First Stars

I n 2010, after completing two decades in orbit and thousands of mind-bending observations, the Hubble Space Telescope (HST) will finally be shut down and returned to Earth aboard a shuttle, for eventual display at the Smithsonian Institution's National Air and Space Museum in Washington.

Future museum-goers will no doubt marvel at Hubble's long list of accomplishments and its spectacular imagery, which establish the $1.5-billion telescope as one of the premier scientific instruments of its time. One seemingly innocuous photograph almost certain to be included shows hundreds of dim galaxies scattered like uncut diamonds on a sheet of black velvet. Casual viewers may give it no more than a brief glance, quickly moving on to visually more dramatic close-ups of star-forming nebulae, exploding suns, and galactic

Floating like an arrow in space, the European Space Agency's XMM-Newton—for X-ray Multi-Mirror—searches for the x-rays that many celestial objects generate but can't be detected from Earth due to our atmosphere. Orbiting since December 1999, XMM-Newton has already examined a cataclysmic binary star, gained insights into black holes, and analyzed the morphology of supernovae remnants.

nuclei harboring super-
massive black holes.

But visitors more
familiar with the observatory's feats will instantly
recognize the Hubble Deep
Field, that mind-bending
photograph of 1,500-plus
galaxies—uncounted tril-
lions of stars—crammed
into what was thought to
be an *empty* area of sky,
about the size of a dime at
a distance of 75 feet.

"The variety of galaxies
we see is amazing," Robert
Williams, director of the
Space Telescope Science
Institute, reflected when
the image was released in
1996. "We are clearly seeing some of the galaxies as
they were more than ten
billion years ago, in the
process of formation."

Our museum-goer,
awed once again by the
Hubble Deep Field, might
pause before moving on, to
wonder just when the first
stars did begin lighting up
in the afterglow of the Big
Bang. She or he might even

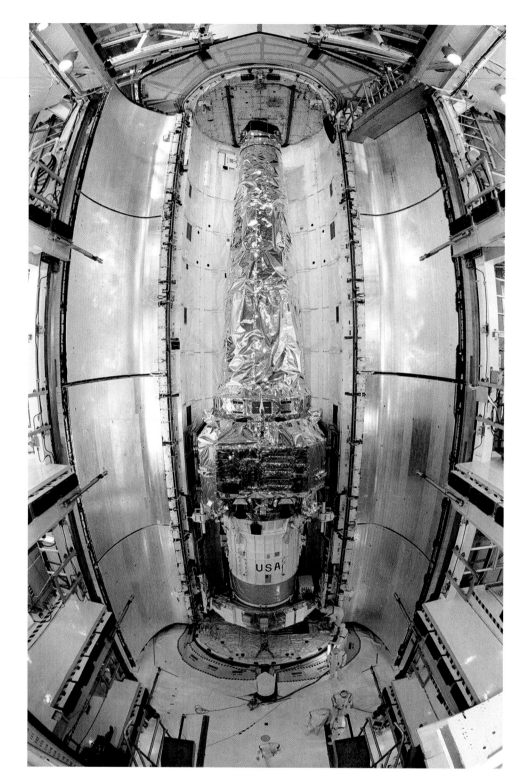

be aware that a million miles from Earth, a new space
telescope launched in 2009 will at that moment be
peering even deeper into space and time, making
observations needed to answer that very question.

At the heart of this Next Generation Space Telescope (NGST) is a segmented mirror measuring 20
feet or so in diameter and packing nearly 6.5 times the
light-collecting area of Hubble's 94.5-inch primary. To
maximize its sensitivity to faint infrared light from the

**Space-age protective blankets shield NASA's Chandra
X-ray Observatory as it enters the cocoon-like payload
bay of space shuttle Columbia prior to liftoff.
Launched into a highly eccentric orbit, Chandra
travels as far as 87,000 miles from Earth—one-third
the way to the Moon—in order to escape our planet's
radiation belts. It studies the physics of various phenomena, including black holes and quasars.**

initial epoch of galaxy formation, NGST will operate in the shade of a sun shield the size of a tennis court. Chilled to 50 degrees above absolute zero, it will be 100 times more sensitive to infrared light than Hubble and 10 times sharper, thus opening a window on the time when stars first took shape.

"What Hubble Deep Field is hinting at is little clumps, little pieces of galaxies starting to form," says Edward Weiler, NASA's associate administrator for Space Science and a former Hubble project scientist. "What we're not seeing is the first generation of stars when those hydrogen clouds first started lighting up. When did the first star in the universe start to burn hydrogen? NGST should be able to tell us."

The 12,700-pound NGST will be launched with a light-splitting infrared spectrometer and two cameras optimized for different infrared wavelengths. Data will be downlinked directly to a dedicated 65-foot antenna on Earth. Unlike Hubble, which operates in low-Earth orbit and is periodically serviced by shuttle astronauts, NGST will function far beyond the reach of astronaut repair crews. It will launch directly to the L2 Lagrangian point, where the gravity of the Earth and Sun are in rough equilibrium. In such gravitational eddies, spacecraft can operate with a minimum expenditure of fuel. The SOHO solar observatory, for example, occupies the L1 point on the other side of Earth, closer to the Sun.

The L2 orbital station offers an unimpeded view of the sky. Also, because the NGST will not be swinging into and out of Earth's shadow every 45 minutes—as Hubble now does—temperature-related stresses will be eliminated, simplifying the design requirements.

"We decided that it might make more sense to go directly to some place like L2 and, if the thing failed, to just launch another one," says Bernard Seery, project manager for NGST. "That would be, in the long run, cheaper than building one to be serviced in low-Earth orbit." Hubble, for example, cost $1.5 billion to build. But throw in the costs of new instrumentation, ground operations, and a half-dozen shuttle flights at roughly $400 million each, and the total price tag over the life of the project swells to some $6.4 billion. In contrast,

the NGST is expected to cost about $1.3 billion, which includes some $500 million for the telescope itself, $200 million or so for technology development, and about $400 million for ground operations and the unmanned rocket that will be needed to get it to L2. As Seery points out, two NGSTs could be built, launched, and deployed—if that became necessary—for less than what NASA has spent on Hubble through the first half of its 20-year mission.

Even so, cost-cutting continues. NASA initially envisioned a 26-foot mirror for NGST, then reduced its size and eliminated an orbital test flight, for economy. Major engineering hurdles also remain. The segmented mirror, while smaller and sturdier than the original design, still must fold up to fit inside the rocket's nose cone for launch. Once in space, it will need to unfold like a flower blossom, using sophisticated sensors and actuators to keep each segment precisely aligned. The sun shade is another challenge: Engineers initially considered an inflatable shield but ultimately opted for a more traditional—and complex—mechanical design.

"Most missions try to bring one or two advanced technologies into the mix," says Seery. "We're trying to bring four or five, depending on how you count. If we hadn't spent about $200 million before launch on technology development, I'd be scared to death."

In the meantime, NASA plans two more shuttle missions to Hubble, to install new solar arrays and instruments. In early 2002, astronauts will attempt to revive an infrared camera and install a new visible-light camera with five times the sensitivity and ten times the efficiency of the telescope's current imaging system.

"That efficiency translates into doing something like the Hubble Deep Field, except deeper," says David Leckrone, Hubble's chief project scientist. "With the Hubble Deep Field, we probed back to maybe when the universe was about a billion years old. Hopefully, we can go back a little bit deeper with the advanced camera and get a little more into NGST territory, using the Hubble Space Telescope as a pathfinder."

NGST also will tackle the most pressing question in modern astronomy—the apparently accelerating expansion of the universe—by searching for Type Ia supernovae in extremely remote galaxies. Such supernovae result when white dwarf stars in binary systems absorb mass from their companions. Once they reach about 1.4 times the mass of our Sun, a wave of catastrophic fusion reactions begins that, within seconds, causes the star to explode.

Because all Type Ia supernovae involve stars of roughly equal mass, the intensity of light emitted by their explosions follows a predictable pattern, brightening to a certain level and then fading away. By comparing the apparent brightness of a supernova with its inferred absolute magnitude, scientists can compute that star's distance from Earth. Thus Type Ia supernovae can be used as "standard candles" to determine distances to galaxies that are beyond the range of other measurement techniques. Traditional spectroscopic analysis tells scientists how fast the supernova and, by extension, its host galaxy are receding from us.

In the late 1990s, astonished researchers discovered that Type Ia supernovae at extreme distances appeared dimmer than expected, based on their recession velocities. The researchers concluded that the expansion of the universe must be accelerating rather than slowing down, as had been believed. They dubbed the unknown force behind this acceleration "dark energy."

In April 2001, Hubble researchers added another piece to this puzzle. Studies of the most distant Type Ia supernova yet observed, which occurred some 10 billion years ago, indicated that the universe's expansion slowed down and then speeded up, as the repulsive dark energy won out over gravity. One discoverer, Adam Riess of the Space Telescope Science Institute, observed, "This supernova shows us the universe is behaving like a driver who slows down approaching a red stoplight and then hits the accelerator when the light turns green." With the installation of Hubble's Advanced Camera for Surveys, scheduled for early 2002, HST is poised to take an even more important role in the ongoing search for Type Ia supernovae.

"We're at the point now where [distant supernovae] are really, really difficult to find from the ground any

more," adds project scientist Leckrone. "This work has to be done in space [and] Hubble will carry the torch."

With Hubble and NGST setting the pace in space-based optical astronomy, NASA, the European Space Agency (ESA) and Japan plan over the next ten years to launch additional satellite observatories that will make use of the entire electromagnetic spectrum. NASA's Hubble-class $1.6-billion Chandra X-ray Observatory, launched in 1999, is expected to operate throughout the decade as well, studying the physics of black holes, quasars, and other high-energy targets. Joining Chandra on the frontiers of x-ray astronomy is ESA's X-ray Multi-Mirror telescope, dubbed XMM-Newton. Also launched in 1999, it is equipped with three x-ray telescopes and five cameras and spectrographs.

In July 2002, NASA will launch the $450-million Space Infrared Telescope Facility, or SIRTF, the most advanced infrared telescope ever. Cooled to just five degrees above absolute zero, its 33-inch telescope will "see" the heat given off by planets and peer through clouds of interstellar and intergalactic dust that block visible light. Targets include distant galaxies, stars in the early stages of their evolution, and protoplanetary disks that may give rise to extrasolar planets.

"The Space Infrared Telescope Facility will do for infrared astronomy what the Hubble Space Telescope has done in its unveiling of the visible universe, and it will do it faster, better, and cheaper than its predecessors," Wesley Huntress, then director of Space Science for NASA, said in 1998 when the project moved into the development phase.

The European Space Agency plans to launch its own satellite to the L2 Lagrangian point in 2007. Equipped with an 11-foot primary mirror, the Herschel Space Observatory will concentrate on galactic evolution and star birth. Meanwhile, NASA and the German Aerospace Center are building an 8.2-foot infrared telescope that will ride aboard a special 747 jumbo jet starting in late 2002. Called the Stratospheric Observatory for Infrared Astronomy, or SOFIA, it will operate at 41,000 feet, above 99 percent of Earth's atmosphere. Its first decade of operation is projected to cost about $310

Taking on some of the biggest questions of the cosmos, NGST—the Next Generation Space Telescope—will observe the formation of the earliest galaxies and stars. What is the shape of the universe? How do galaxies evolve? How do stars and planetary systems form and interact? NGST will probe such challenges after its launch, planned for around 2009.

million. While space-based telescopes can observe the full range of infrared wavelengths, SOFIA will be much more easily upgraded with state-of-the-art instruments, thus it should have a longer operational lifetime. SOFIA is expected to operate for 20 years.

In addition to Hubble, Chandra, and SIRTF, NASA's original four "great observatories" include the $670-million Compton Gamma-Ray Observatory (CGRO), launched in 1991. Originally designed to operate for just five years, CGRO ended up working for nearly ten before crashing back to Earth, in June 2000. Over its orbital lifetime, it discovered 70 gamma-ray quasars, 10 rare gamma-ray pulsars, and more than 2,500 gamma-ray bursts, the most titanic explosions since the Big Bang.

No one knows for sure what causes these short-lived explosions. They may be related to black hole formation, to collisions between neutron stars, or to even more exotic phenomena. In any case, studying

them in greater detail remains a high priority for astronomers around the world. ESA plans to launch the International Gamma-Ray Astrophysics Laboratory, or INTEGRAL, spacecraft in 2002; NASA hopes to launch Compton's successor, the Gamma-Ray Large Area Telescope, or GLAST, around 2006. To help resolve the mystery of gamma-ray bursts, NASA also plans to launch the smaller Swift Gamma Ray Explorer, in 2003. Equipped with a trio of gamma-ray, x-ray, and optical telescopes, the $163-million Swift can be aimed at a gamma-ray burst within minutes of its initial appearance. It also will search for black holes and other gamma-ray sources.

Two other spacecraft, launched by NASA and ESA, will probe the faint heat left over from the Big Bang itself. The goal is to determine how minute differences in density gave rise to the "lumpy" universe of stars and galaxies we see today. NASA's Cosmic Background Explorer satellite (COBE) conducted the first space-based study of that leftover heat, measuring background radiation that is detectable throughout the universe. Its four-year mission ended in 1993, after it had found an unevenness, known as anisotropy, in the background radiation amounting to one part in 100,000. But COBE could not distinguish variations across areas of the sky smaller than about seven degrees.

NASA's $145-million Microwave Anisotropy Probe (MAP), just launched in June 2001, will spend 27 months at the L2 Lagrangian point charting variations across areas just two-tenths of a degree across. That's a 15-fold improvement over COBE. And in 2007, the European Space Agency plans to launch Planck, a spacecraft named in honor of German physicist Max Planck, which will map the background radiation of the universe at even higher resolution.

Observes MAP principal investigator Charles Bennett, "The miracle here is that there is a light from the early universe and that it carries all this information with it and that we can measure it. I find that astounding. It's not obvious that humans on the planet Earth should be able to understand the whole universe. But that appears to be the path we're on."

Fountain of annihilating radiation—produced by the interaction of matter and antimatter—points up, away from the plane of our galaxy. Its source is the galactic center, shown here as the bright red circular region. When a particle and its antiparticle—such as an electron and a positron—meet, they annihilate each other, producing neutral pions, which quickly decay into gamma-rays. The Compton Gamma-Ray Observatory, launched in 1991, gathered the data for this map; the colors represent the intensity of gamma-ray emission from positron-electron annihilation. An overlay of radio contours suggests that a channel leads from the center of the Milky Way to high latitudes, in general agreement with the location and direction of the annihilation fountain.

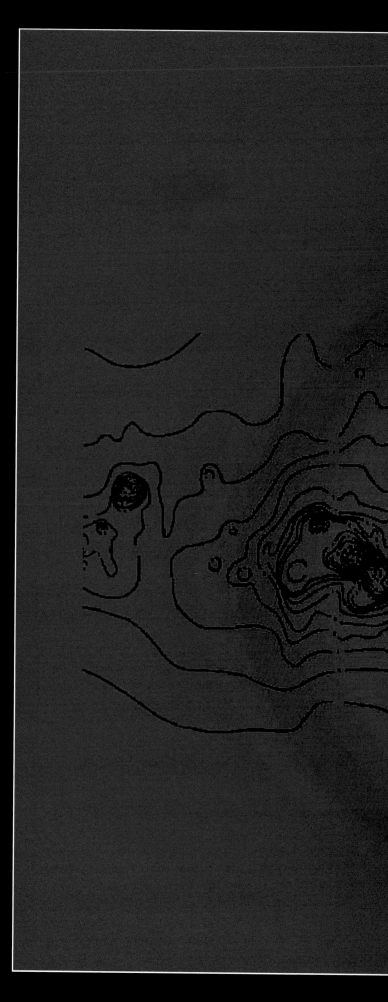

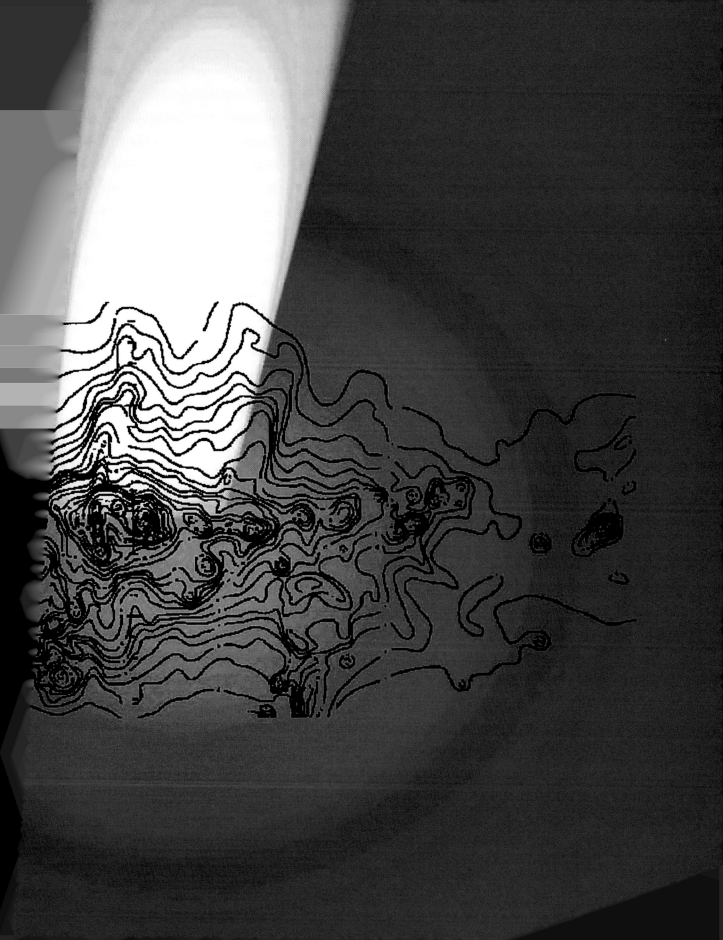

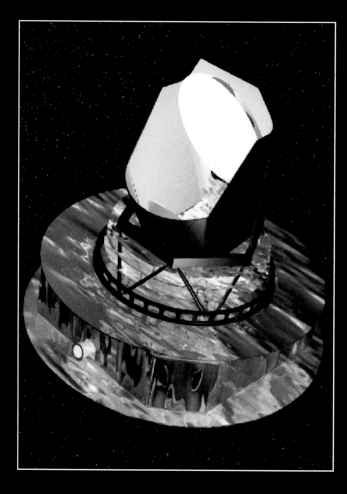

Due to launch in 2007, the European Space Agency's Planck satellite (right) will tally radiation from shortly after the Big Bang, radiation that is still surfing the universe like a shock wave from the event itself. Called CMB, or Cosmic Microwave Background, this radiation holds data about the initial conditions that gave rise to today's visible universe, over the past 10 to 14 billion years. To read that data, Planck will measure the temperature of the entire sky and chart small fluctuations. An all-sky map similar to this simulation (below) will result, showing individual hot and cold spots, each comparable in size to clusters and superclusters of galaxies.

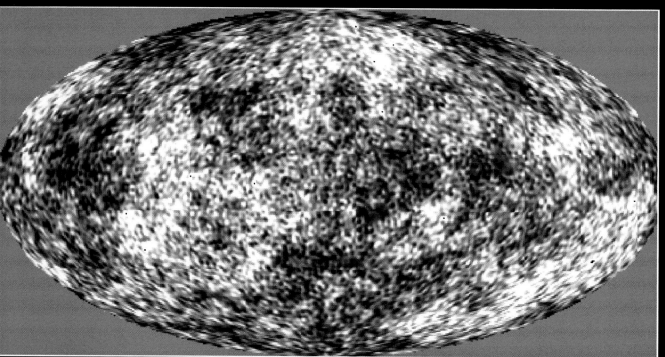

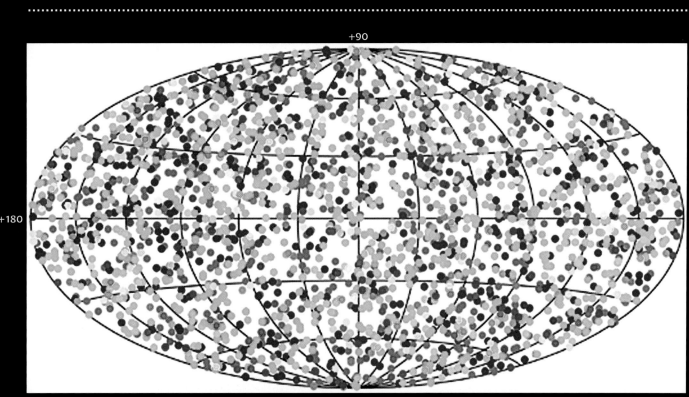

Sizzling eruption of antimatter spews from the center of the Milky Way (above), as recorded by the orbiting Compton Gamma-Ray Observatory, or CGRO. The blast could be a result of massive numbers of stars being born near a large black hole at the center of our home galaxy. Discovering it, says one astronomer, "is like finding a new room in the house we have lived in since childhood. And the room is not empty...." Launched into Earth orbit from space shuttle Atlantis in 1991, CGRO served until a failed gyroscope forced NASA to destroy it in June 2000. Before its demise, it recorded the positions of gamma-ray bursts in the sky (below); colors denote the brightness of individual bursts. The telescope detected more than 400 gamma-ray sources and 2,500 gamma-ray bursts, in both cases about 10 times more than had been known previously. Astronomers are still trying to determine the sources and causes of gamma-ray bursts.

+90

+180

-90

MAKING BIG TELESCOPES FROM SMALL

The Magic of Interferometry

When it comes to telescopes, size matters—and bigger is definitely better. Even if the telescope exists only in the virtual world of a computer's central processor. At the W. M. Keck Observatory atop Hawaii's dormant Mauna Kea volcano, astronomers are combining light from two 33-foot telescopes to produce a simulated telescope with the resolving power of a mammoth 279-foot mirror. Such a virtual giant would be capable of distinguishing, in theory at least, a beach towel on the surface of the Moon. Or a Jupiter-size planet orbiting another star.

At the European Southern Observatory's aptly named Very Large Telescope (VLT) facility in northern Chile, astronomers are on the verge of combining the light from four already operational 8-meter (26.2- foot) telescopes to produce a truly gargantuan virtual mirror. On Mount Graham near Safford, Arizona, an

Earthbound or orbiting, optical or radio, telescopes of tomorrow will be bigger, more complex, and better at unlocking secrets of the universe. The 6.5-meter (21.3 foot) Multiple Mirror Telescope, designed for imaging and spectrographic studies of stars and galaxies— looks out from Arizona's Mount Hopkins.

international consortium is completing a giant set of binoculars with a pair of 27.6-foot mirrors that will have the light-gathering power of a 39-foot telescope and the resolving power of a single 75-foot mirror. The scientific sleight of hand that makes this possible is optical interferometry, a technique that utilizes state-of-the-art technology and the wave nature of light to achieve high angular resolution—the ability to distinguish separate objects—at one-tenth or one-twentieth of the cost of a comparable full-size telescope.

Radio astronomers have used interferometry for decades, building arrays of dish antennas and bringing together their signals in a precisely controlled fashion. By measuring the interference patterns that result as radio waves reinforce or cancel each other out, computers assemble an image that simulates the output of a single huge antenna. The technology needed to combine radio waves is relatively straightforward. But the technology needed to successfully combine much shorter wavelengths, which include infrared and visible light, is fiendishly complex. To achieve the interference patterns needed to reconstruct an image, the light waves from multiple telescopes must arrive precisely in step at the beam combiner, achieving an accuracy

Eyes wide open, the twin domes of Hawaii's Keck Observatory each rely on a 10-meter (33-foot) primary mirror, composed of 36 hexagonal segments. Computer-controlled sensors adjust the mirrors to an accuracy of four nanometers—1,000 times thinner than a human hair—to counteract the tug of gravity.

of—at worst—about a thousandth of a millimeter, or 0.0000394 inches. For large telescopes separated by several hundred yards, the actual beam paths must match even more closely.

Ground-based interferometers also face another challenge: atmospheric turbulence. To ensure the light beams stay synchronized, computer-controlled feedback loops are required to constantly fine-tune each telescope's optical system to exactly counteract the effects of the atmosphere on incoming light. Technology to accomplish this first became widely available in the 1990s. Now, optical interferometry is emerging as a cost-effective way to greatly increase the reach of even the largest ground-based optical telescopes.

"If you want a certain collecting area, it's cheaper to build a bunch of small telescopes rather than one large one," says Michael Shao, director of the Interferometry Center for Excellence at the Jet Propulsion Laboratory (JPL). "As telescopes get bigger and bigger, the dome gets bigger, it's more massive, the mechanical structures get harder to build. The radio folks have basically gone to interferometers mainly for economic reasons and I think, ultimately, that'll be the same reason optical folks go that way." He adds that, in the 1990s, "a number of interferometers were built with small telescopes, basically to prove the technique. And this decade is when that technology is going to be applied to the largest telescopes on the planet."

For more than 45 years, the mighty 200-inch Hale Telescope atop Mount Palomar in southern California stood as the ultimate achievement of the telescope maker's art, a massively elegant marriage of precision optics and brute-force technology that brooked no rival. Completed in 1947 and still operational, the telescope's nearly flawless mirror is housed in an open-framework tube carried by a superbly balanced equatorial mounting that tips the scales at 530 tons. Its huge rotating dome weighs nearly twice that. Although the Soviet Union commissioned a 236-inch telescope in 1976, it suffers optical defects, a relatively poor viewing location, and other technical shortcomings. As a result, Palomar remained the Mount Olympus of the astronomical community throughout the 1980s and early 1990s, when advances in lightweight materials, adaptive optics (to counteract atmospheric turbulence), and improved techniques for casting mirrors sparked a telescope building boom that continues to this day.

In 1993, the first of the twin 33-foot Keck telescopes atop Mauna Kea came on line, far surpassing Hale. They remain the world's largest, although instruments with mirrors up to three times larger already are in preliminary planning stages. European researchers are even discussing the feasibility of a truly gargantuan 330-foot telescope. But for now, Keck is king and interferometry is the name of the game. Once Keck's two main telescopes and its four 6-foot "outrigger" scopes become fully operational as an interferometer, it should be able to see so-called "hot Jupiters" orbiting close to nearby suns, and it will indirectly detect Uranus-size planets around stars up to 60 light-years away.

The first planets found outside of our solar system, Shao explains, were Jupiter-size planets that orbited "very, very close" to their stars. "Because they're so close, they're actually quite hot—on the order of a thousand kelvins [1,340°F]. Our Earth in visible light is ten billion times fainter than the Sun; a hot Jupiter is only ten thousand times fainter than the star it's sitting next to. So it's a million times easier to see a hot Jupiter than it is to see an Earth. That factor of a million is big enough that we hope, a couple of years from now with the Keck interferometer, we'll be able to see the light from that planet."

Researchers may even use Keck to take spectra of the atmospheres of extrasolar planets to determine their basic chemical composition, thus gaining profound new insights into the structure and environment of such bodies. Chile's VLT interferometer and others will extend that reach even more and help scientists lay the groundwork for space-based interferometers capable of detecting Earth-size planets and even analyzing their atmospheres for signs of biological activity.

"For 2,000 years people have been asking these questions," says Charles Beichman, program scientist with NASA's Origins program at JPL. "The big thing is we now have the technological tools at hand to make that not a philosophical question but a scientific question. We already know there are other planets out there. Within ten years we'll know if there are Earth-like planets out there and, in 20 years, we should know if any of those Earth-like planets support life."

NASA is funding the Keck interferometer project as a pathfinder to develop techniques and technology needed to design, build, and launch the first space-based optical interferometer in 2009. The $930-million Space Interferometry Mission (SIM) spacecraft will be stationed in solar orbit for a five- to ten-year mission. It will use interferometry to determine the precise positions of hundreds of stars and to measure the minute wobbles they experience due to the gravity of relatively large planets that are orbiting them.

"With SIM, you see the position of a star wobbling with respect to distant background stars," says Beichman. "So what you really have is a little map which shows some circular gyrations of a star that you then turn into a computer model and ask, what planet or planets would produce those gyrations? And from that, you solve Kepler's equations and infer that there's this planet at that distance in this orbit and so on."

Anne Kinney, director of NASA's Astronomy and Physics division, adds, "Say you were 10 parsecs [33 light-years] from the Sun. If you were to observe the Sun with high accuracy, you would see it moving around in the sky as Jupiter tugs on it. If you were to see it with extremely high accuracy, those movements would have little tiny sinusoids on them as the Earth goes around (the Sun), as Mars goes around it, and so on. That gives you the ultimate solution for an entire solar system. What SIM allows us to do is view exactly what sits around every star in our neighborhood. Does each one of them have a solar system, a Jupiter, a Saturn, a Mars? Or are none of them like us at all?"

A cosmic census-taker, SIM will examine several thousand relatively nearby stars and catalog their planets that are at least ten times more massive than Earth. It then will study a few hundred stars in enormous detail to find planets as small as three Earth masses. The SIM spacecraft features three co-aligned interferometers mounted on a 33-foot-long boom. Each combines the light from two telescopes. Two of them will lock onto guide stars to keep the spacecraft precisely oriented, while the third examines target stars and planets. SIM's virtual mirror will have the resolution of a single 33-foot telescope.

Although atmospheric distortion will not be a factor, the job remains enormously complex. Light paths between SIM's mirrors, for example, must be actively stabilized to nanometer—a billionth of a meter, or 0.0000000394 inch—precision. Mirror positions must be accurate to within the width of a few hydrogen atoms. Says Kinney, "It has to have nanometer control of the optical elements and picometer [trillionth of a meter] sensing of the elements. We have achieved the

nanometer part, we are working on the picometer part. It is, as you can imagine, non-trivial."

Technology that permits such precision is crucial as well for a follow-up to SIM: The multi-satellite Terrestrial Planet Finder (TPF), a set of five free-flying interferometers capable of directly imaging Earth-size planets orbiting nearby stars and taking spectra of their atmospheres to look for signs of biological—or industrial—activity. The TPF program is expected to cost up to $2 billion. Launch is targeted for around 2012, although Kinney says that date will be tough to meet, because its spacecraft and instrument designs are still in preliminary stages.

To accomplish its remarkable goal, TPF will utilize interferometry to cancel the light from a star in order to see the starlight reflected from its much dimmer planets. "At that point," says Beichman, "you're actually starting to use remote-sensing techniques to characterize the atmosphere of the planet. And in some fraction of the cases, you might find Earth-like signatures."

The search for Earth-like planets is a major driver in the development of space-based optical interferometry. But for sheer scale, the radio astronomers who pioneered astronomical interferometry still rule the roost. A consortium of U.S. universities and organizations is studying concepts for a proposed $600-million array of more than 100 antennas, with a total collecting

Chile's high-altitude desert awaits the arrival of
ALMA—the Atacama Large Millimeter Array.
Sixty-four dish antennas, each 39 feet across, will
sprout from the dry Zona de Chajnantor, elevation
16,400 feet. At far left, the perfect cone of Licancabur
volcano rises near the Bolivian border.

area of one square kilometer (247 acres). The National
Science Foundation plans to expand the Very Large
Array (VLA) in New Mexico from 27 to 35 moveable
dishes, improving its overall resolution tenfold, for a
relatively modest $140 million.

At shorter wavelengths, U.S., European, and Japanese researchers hope to complete the $400-million
Atacama Large Millimeter Array (ALMA) in Chile by the
end of the decade. This will include 64 dish antennas,
each 39 feet in diameter. According to the proceedings
of a 1995 European Southern Observatory workshop,
ALMA will "detect and study the earliest and most
distant galaxies, the epoch of the first light in the
Universe. It will also look deep into the dust-obscured
regions where stars are born to examine the details of
star and planet formation."

Space-based interferometry at radio wavelengths
already is a reality: The international Very Long Baseline Interferometry Space Observatory Program (VSOP)
utilizes ground-based antennas as well as a satellite
antenna to create a huge virtual antenna. Explains
David Murphy, a researcher at JPL, "Because we go to
space, we have a longer baseline, therefore we can get
higher resolutions. For a given observing frequency,
our linear resolution is three times better than anything
that can be obtained on the Earth."

The Japanese Institute of Space and Astronautical
Science hopes to launch a follow-up VSOP spacecraft
around 2008. Beyond that, Murphy says, "we're thinking of multiple-telescope missions. We're thinking
about bigger antennas in space. We're thinking of
going to higher frequencies and higher data rates."
Sensitivity to higher frequencies is crucial to studying
some of the most violent objects in the universe, from
quasars and active galactic nuclei to the magnetically
driven jets that occur near black holes.

"We're trying to figure out how those jets are accelerated, we're trying to work out a lot of jet physics,"
Murphy says. "There are various shock structures,
there are magnetic field structures. We're trying to
figure out the properties of the material near the black
hole." He adds that "the objects we look at are also the
same objects that are emitting lots of gamma rays and
lots of x-rays. So there's a very good tie-in between
our subject and high-energy astrophysics, and the
next-generation high-energy astrophysics missions."

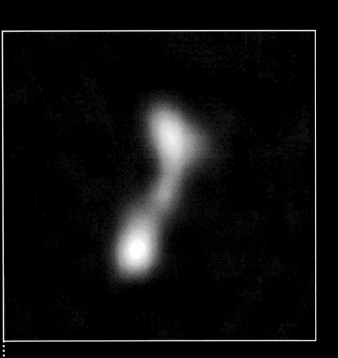

A three-minute exposure from FIRST—Faint Images of the Radio Sky at Twenty centimeters—(above) is part of a project to produce the radio equivalent of Palomar Observatory's optical Sky Survey. The survey is being conducted at the Very Large Array (VLA), one of the world's premier astronomical radio observatories—27 antennas, each 82 feet in diameter, arrayed in a huge Y pattern across the New Mexico desert.

When completed, ALMA's huge dish antennas (right) will focus on millimeter and sub-millimeter wavelengths, between radio and infrared spectral regions. Such observations are essential to understanding the star-forming history of the universe, astronomers say. A collaborative effort among the U.S., Japan, and several European nations, ALMA will be one of the largest ground-based astronomy projects ever.

Fast, but not fast enough: Receding from Earth at some 1,200 miles a second—and already some 100 million light-years away—the galaxy NGC 6070 (above) can't escape the watchful eye of the Sloan Digital Sky Survey telescope (right). Blue points of light come from recently formed hot stars. The most ambitious astronomical survey ever undertaken, the Sloan effort will systematically map a quarter of the entire sky, determining the position and absolute brightness of more than 100 million celestial objects. By measuring distances to a million of the nearest galaxies, it will produce a three-dimensional picture of this corner of the universe.

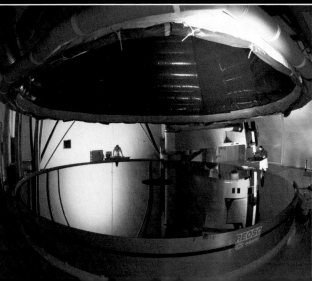

Atop Mt. Paranal in the driest part of Chile's Atacama Desert (above), four structures housing the aptly named Very Large Telescope (VLT) near completion. These four and several smaller telescopes will be used in combination as a giant interferometer—an instrument for precisely measuring interference patterns produced when radio waves either reinforce or cancel each other out; together, they will act as a single gigantic antenna. One of the four 8.2-meter (27-foot) main mirrors undergoes final polishing (left). The VLT will be able to detect astronomical objects ten billion times fainter than can be seen with the naked eye.

PART III

THE NEXT CENTURY

TECHNOLOGY LEADS THE WAY

Propulsion Systems, from Ion Drives to Antimatter

T wenty-four middle-class American males share the solar system's current speed record for manned flight: 25,000 mph, or 6.9 miles per second. That's how fast a three-seat Apollo moonship had to go to break free of Earth's gravity. No human has ever moved faster.

But given that our solar system is some 7.3 billion miles across, 25,000 mph is a slow crawl. At that speed—fast enough to circle the entire globe in an hour—it would take more than 16 years to reach Pluto, assuming our intrepid astronauts maintained a constant velocity. And one would need 118,000 years to reach the nearest star, 25.9 trillion miles away.

The core problem is that Apollo moonships and virtually all other major rocket systems yet devised rely on chemical energy for propulsion, burning a fuel with an oxidizer to produce

Tomorrow's technology today: This solar-powered ion engine—the first non-chemical propulsion system used in a spacecraft—powers Deep Space 1, a probe launched in October 1998. DS-1 tested this and a dozen other new technologies, including onboard software that tracks celestial bodies so a spacecraft can make its own navigational decisions, without the help of ground controllers.

thrust, much the way an automobile engine burns gasoline with atmospheric oxygen to turn a drive shaft. It works, but it's terribly inefficient.

Aside from the velocity boosts available from gravity through close planetary flybys, the only practical way to significantly shorten trip times for chemical rockets is to decrease the mass of the payload or carry more fuel. The first option decreases a mission's scientific payoff; the second sharply increases the size, cost, and complexity of the launch vehicle. For nearly 40 years, mission planners have faced this conundrum in one form or another and have struggled to maximize the science return of a given space mission while minimizing trip time and overall cost.

But on October 24, 1998 that equation changed for the first time since the dawn of the space age, with launch of a small experimental spacecraft called Deep Space 1. While the $150-million DS-1 rode into space atop a traditional chemical-fuel rocket, the spacecraft itself was equipped with a much more efficient solar electric ion drive. Instead of burning fuel and oxidizer in a chemical reaction, its ion drive uses electrical power from solar cells to strip electrons from the atoms in the spacecraft's propellant, leaving positively charged ions. An electric field then accelerates those ions to enormous velocities—nearly 70,000 miles an hour aboard DS-1—and then directs them out the back of the rocket to produce thrust.

While the thrust of an ion rocket is relatively gentle—the flow rate of ions is vanishingly small—the engine is so efficient that it can fire continuously for years using a small amount of propellant to slowly but surely boost the spacecraft's velocity beyond anything achievable by a chemical rocket.

The standard yardstick for measuring a rocket engine's efficiency is specific impulse, the number of seconds an engine can produce one pound of push while consuming one pound of fuel. The space shuttle's hydrogen-powered main engine, the most efficient chemical rocket engine currently in service, has a specific impulse of slightly more than 450 seconds. That is roughly seven to ten times less than the specific

impulse of the DS-1 ion drive. Put another way, the ion drive on Deep Space 1 can fire continuously for 20 months—and use up just 180 pounds of xenon fuel. In the process, its velocity will rise by nearly 10,000 mph. The same amount of fuel in the most efficient chemical rocket produces a velocity boost of less than 1,000 mph.

While ion engines have been studied for decades in various forms, the DS-1 spacecraft was the first to rely exclusively on ion propulsion, and to prove the technology's feasibility in the actual environment of space. At the Johnson Space Center in Houston, physicist and astronaut Franklin Chang-Diaz and a team of scientists and engineers are developing a more sophisticated version of the solar electric ion drive called the variable specific impulse magnetoplasma rocket, or VASIMR.

"What we do," says systems engineer Andrew Petros, "is use radio frequency energy to ionize a gas and then further heat that gas. Once the gas is ionized and becomes a plasma [a fourth, high-energy state of matter] it can then follow magnetic field lines. We create a magnetic field around the whole device with a series of coils, electromagnets. It's a magnetic bottle open at one end, so the plasma will come out the open end at high velocity."

Unlike the DS-1 ion thruster, the VASIMR engine does not use electrodes to accelerate the propellant ions, nor does it need to inject electrons into the exhaust flow to neutralize the spacecraft's electrical charge. It also uses a lighter propellant. While such gases are more difficult to ionize, they can be accelerated to higher velocities, Petros says. And unlike the DS-1 thruster, the VASIMR's specific impulse can be varied in flight to maximize performance. In addition, he says, VASIMR "should be more durable, it shouldn't have anything that will degrade over time or erode. The ISP [specific impulse] should be very high."

In theory, up to 30,000 seconds. Small-scale prototypes using 10 to 25 kilowatts of electrical power are expected to have specific impulses ranging from 5,000 to 10,000 seconds. The VASIMR team hopes to mount a prototype thruster on the international space station around 2004 to demonstrate the system's feasibility.

With a splash and a roar, Cosmos 1 lifts off atop a Volna rocket from a Russian submarine in this rendition of the first solar sail mission, a project envisioned by the Planetary Society and Cosmos Studios, both private organizations co-founded by astronomer Carl Sagan. Like an old-fashioned square-rigger, the fuel-free craft will be powered through space not by wind but by reflected light pressure pushing against giant sails. In Sagan's words, "We have lingered long enough on the shores of the cosmic ocean. We are ready at last to set sail for the stars."

If it works as expected, NASA potentially could reduce or even eliminate costly unmanned flights to deliver rocket fuel to the outpost. "They're going to be launching tons of propellant to the station over its lifetime to keep it from falling out of orbit," Petros says. "We can save them a tremendous amount of propellant at the expense of some more electrical power."

But the real payoff will come with solar system exploration. Petros says the VASIMR rocket can be scaled up to support human missions to Mars or to send probes on high-speed direct voyages to virtually any other planet in the solar system. Unlike chemical rocket missions to the outer planets, spacecraft with ion drives like VASIMR could fly directly to their targets, reducing trip times in part by eliminating the need for

gravity-assist flybys of planets that may or may not be along the way. With enough electrical power, such craft would even be able to brake into orbit around an outer planet, a feat that is extraordinarily difficult to accomplish with chemical rockets.

But robust, high-speed missions like these assume the use of multi-megawatt power plants to provide the electricity needed for life-support systems, instrumentation, communications, and the drive itself. And the only way to achieve such power levels in the foreseeable future is through the use of nuclear reactors.

"If we want to do these deep space missions at all, we'll consider other power sources," Petros maintains. "There's some work being done on more advanced solar power systems, which will be great for near-Earth exploration. But you can't do much about the Sun's power dropping off as you go out."

While DS-1-class ion drives and the VASIMR engine reflect known physics and relatively mature technologies, other propulsion systems also are under investigation. Robert Winglee, at the University of Washington, is developing an innovative engine in which an electrically charged plasma can be used to inflate the magnetic field of a spacecraft to a diameter of 25 miles or so, creating what amounts to its own mini-magnetosphere. The tenuous but swift solar wind then pushes against this artificial magnetosphere and moves the spacecraft, just as wind propels a sailboat. A specific impulse of up to 30,000 seconds is possible in theory—more than enough to drive a moderate-size spacecraft to Saturn in five and a half years and, once there, get it into orbit.

"You can scale it up to larger and larger spacecraft, but you will need corresponding increases in electrical power," Winglee says. A drive powerful enough for a manned Mars mission, for example, would need the kind of electrical power produced by the international space station's huge solar arrays. With nuclear power, Winglee adds, "we can do a heck of a lot more."

While compact nuclear reactors designed for spacecraft could be available within a few years, anti-nuclear activists promise concerted opposition to any missions that include launching radioactive material on rockets, which occasionally blow up. Such matters are beyond the scope of scientists exploring alternative propulsion technologies, but almost all say privately they support the development of nuclear power systems for interplanetary spacecraft, arguing that the benefits far outweigh the risks.

For a possible mission to Pluto, the benefits are obvious. Regardless of the propulsion technique employed, the spacecraft will need electrical power to carry out its scientific investigations and beam its data to Earth. Current solar cell technology cannot generate enough power beyond the orbit of Mars to be scientifically viable. Past deep-space missions, such as the Voyager probes to the outer planets, relied on radioisotope thermoelectric generators (RTGs), devices that convert heat produced by the decay of highly radioactive plutonium 238 dioxide directly into electricity. RTGs have proven extremely reliable, but their electrical output typically is limited to a few hundred watts at launch— and it drops off over time, as the plutonium 238 decays and only a portion of it remains available for use by the spacecraft's science instruments.

A nuclear reactor, coupled with a high-efficiency ion drive, would provide enough electrical power and propulsion to revolutionize planetary exploration.

"For destinations in the outer solar system, say Pluto, you now have—in addition to the maneuvering capability offered by a high-performance propulsion system—an energy source capable of providing power levels several orders of magnitude greater than current sources," says George Schmidt, deputy director of NASA's propulsion research center at the Marshall Spaceflight Center in Huntsville, Alabama. "This translates into the ability to support much more sophisticated experiments and greatly improved communications at the destination planet. Instead of being limited to instruments in the 10- to 100-watt range, scientists can now think about applications involving

Like a giant pair of binoculars, the two 10-meter (33-foot) main telescopes of Hawaii's W.M. Keck Observatory peer heavenward. Alone, either qualifies as the world's largest optical telescope. Together, they will be the equivalent of a 90-meter (295-foot) instrument, enabling astronomers to study faint and distant objects such as extrasolar planets.

several tens of kilowatts. This would enable detailed surface mappings via radar, along with possible study of internal planetary characteristics."

A nuclear-electric spacecraft could not only achieve orbit around Pluto, it could depart after completing its primary mission and possibly explore one or more Kuiper Belt objects. If costs were no object and political opposition could be overcome, such power plants could be operational "within seven to eight years," Schmidt says. "Definitely within a decade."

Anticipating objections from the anti-nuclear lobby, engineers are designing possible space reactors with safety in mind, from the ground up. For starters, they plan on using enriched uranium 235 instead of the much more radioactive plutonium 238 found in RTGs. To prevent any chance of a reactor core reaching critical mass and exploding—even in the worst-case scenario of a failure occurring within seconds of liftoff—Schmidt suggests, "We would remove a portion of the fuel from the reactor and block it. We call it 'in-space fueling.'" By physically blocking part of the fuel from the core during launch and then activating it later through some mechanical means, he adds,

"We're completely removing the possibility of any inadvertent startup during launch."

Another consideration: Because it offers so much power, a nuclear-electric spacecraft would not need to make a flyby of Earth; this eliminates any chance that its onboard reactor could inadvertently release radiation into our planet's atmosphere. In contrast, chemically propelled missions such as Galileo and Cassini have had to make multiple flybys of Venus and Earth before departing the inner solar system.

One innovative propulsion technology that promises relatively high specific impulse without nuclear power involves solar sails. The idea is to unfurl a huge but lightweight sail that measures miles across, catch the solar wind—or even the incredibly slight pressure of the Sun's light—and use it to propel the spacecraft.

"Sails have a lot of interesting prospects, especially for robotic exploration and fast planetary flybys," says Schmidt. "The thing that's appealing about these is, if you can build them, you can achieve extremely high velocities because even though the thrust is extremely low, it can operate for extremely long periods of time. So you can accumulate quite a substantial velocity."

To bring a solar sail up to suitable cruising velocity, however, it must first pass close by the Sun, Schmidt cautions. This raises a serious technical challenge: Not only does the sail need to be extremely lightweight, it also must be extremely heat resistant. No currently known materials fit that bill.

In the meantime, engineers have not ruled out developing engines that rely on the direct heating of a propellant by a nuclear reactor to achieve less lofty goals. In such a nuclear thermal rocket, a propellant such as hydrogen is pumped into the core of a reactor, heated to very high temperatures, and then expelled out the rear of the rocket. The low molecular weight of hydrogen translates into a high exhaust velocity. Even so, the specific impulse of this type of rocket motor is on the order of 1,000 seconds, only about twice that of current chemical rockets.

A much more attractive long-range option is fusion power, which could produce a specific impulse as high as 150,000 seconds or more. Even though a working fusion reactor appears decades away, the benefits are so great that this technology—if it ever becomes a reality on Earth—no doubt will be adapted for space travel.

"Within the next 50 to 100 years," Schmidt says, "the likelihood of developing devices capable of supporting controlled, high-gain fusion reactions looks good. The extremely high energy densities offered by nuclear fusion could translate to extremely high specific impulses, probably on the order of 50,000 seconds and perhaps several 100,000 seconds. We can really do quite a bit with that type of performance, and because of fusion's extremely high power density as well as its specific impulse, we can now entertain conducting very rapid transportation throughout the solar system. With such technology, transits to Mars could be completed within several weeks."

The energy released from a given amount of fuel in a fusion reaction, he adds, is roughly 8 billion times greater than the energy released in a chemical rocket. That amount of energy is also an order of magnitude greater than what a fission-based engine will produce.

But even fusion power pales in comparison to the energy available from a motor that uses the mutual annihilation of matter and antimatter. If just half of this annihilation energy could be transformed into directed exhaust, specific impulses greater than 1,000,000 seconds would be feasible. The working "fuel" for such a rocket likely would be protons and antiprotons. At present, however, the only way to produce antiprotons is in huge particle accelerators that smash high-energy protons into metal targets. In this process, antiprotons are produced with 0.5 percent efficiency, according to *The Starflight Handbook*, by Eugene Mallove and Gregory Matloff.

All told, only a tiny fraction of an ounce of antiprotons is produced each year in the United States and Europe, at a cost of tens of billions of dollars. Until engineers develop a more economical way to produce antiprotons—not to mention the technology needed to manipulate them in large quantities—antimatter drives will remain a distant dream.

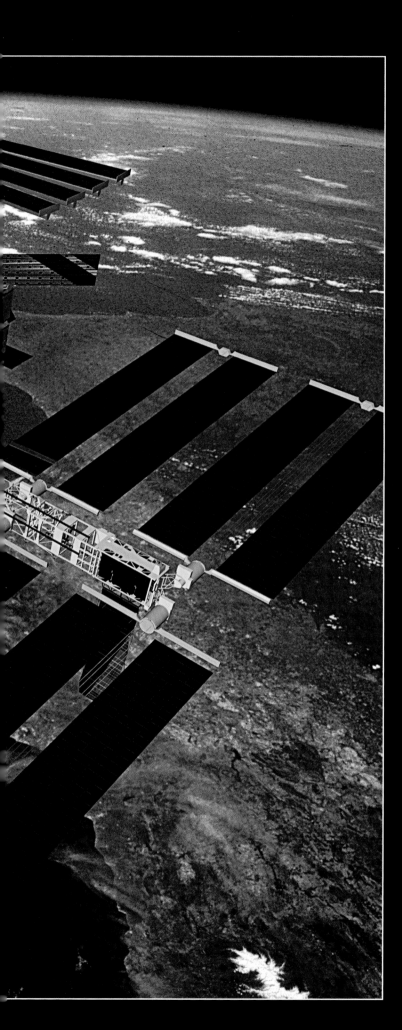

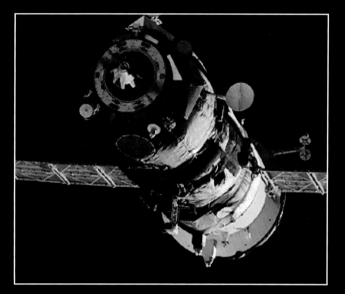

Poised high above the Straits of Gibraltar and the Mediterranean, the International Space Station (left) offers a permanent habitat in space. It is shown as it will appear after completion later this decade. Modules and equipment will be provided by 16 nations, including Japan, Canada, Russia, and several western European countries. Nearly an acre of solar panels will power the structure, which would weigh about a million pounds on Earth. An unmanned supply ship (above) nears the station in November 2000 to dock and deliver its cargo: two tons of food, clothing, hardware, and gifts for the station's crew.

....

Following Pages: Bigger inside than it seems, the completed International Space Station will have a combined interior space roughly as large as two 747 jumbo jets. Multinational crews of up to seven people will live and work here for up to six months.

....

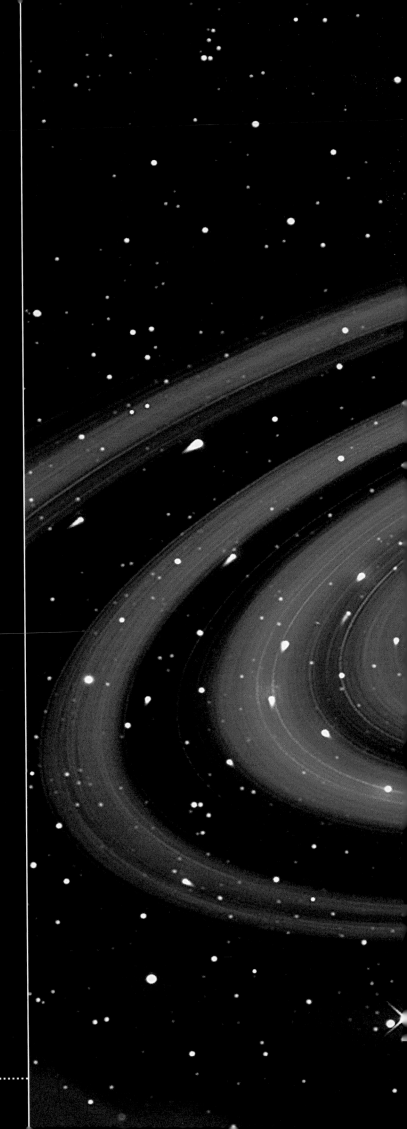

SEEING THE UNSEEABLE

Black Holes and Supersize Telescopes

I n their final moments before departing the observable universe, atoms spiralling into the bottomless pit of a black hole emit blasts of high-energy x-rays, cosmic shrieks in the night that illuminate the boundary between known physics and profound mystery.

Directly detecting and measuring such x-rays would enable scientists to decipher hidden clues about the warped environment just outside a black hole, where Einstein's theory of gravity is stretched to the breaking point and the nature of space-time itself dissolves into uncertainty. But imaging even a giant, extra-galactic black hole's immediate vicinity is an incredibly tall order, requiring an x-ray telescope of mind-bending proportions. It would need one million times sharper vision than the Hubble Space Telescope. That's because resolving a black hole—even one as

Like a celestial whirlpool, this artist's rendition of a rapidly rotating black hole—so named because its gravitational power is so enormous that not even light can escape its clutches—sucks nearby gas and dust into a spiraling accretion disk. As matter falls in toward the center, it is intensely compressed and heated, causing it to emit x-rays. This high-energy glow is evidence of the otherwise invisible black hole.

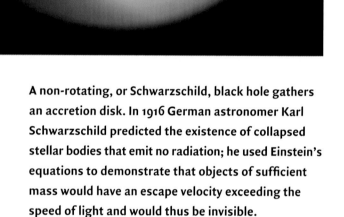

A non-rotating, or Schwarzschild, black hole gathers an accretion disk. In 1916 German astronomer Karl Schwarzschild predicted the existence of collapsed stellar bodies that emit no radiation; he used Einstein's equations to demonstrate that objects of sufficient mass would have an escape velocity exceeding the speed of light and would thus be invisible.

big as our solar system, lurking at the heart of a remote galaxy—is roughly equivalent to discerning a dinner plate on the surface of the Sun from Earth, 93 million miles away.

But that is precisely what a team of far-thinking scientists and engineers proposes to do in a feat of macro- and micro-engineering truly worthy of the 21st century: Put a 34-spacecraft x-ray interferometer in orbit around the Sun and operate it as a gigantic virtual telescope. Thirty-two spacecraft would be arranged in a ring, separated from each other by up to a half mile or so and held together with a complex laser guidance system that keeps them precisely aligned within tiny fractions of an inch. Incoming x-ray photons from the distant black hole under study would bounce off spacecraft mirrors and meet in a special, beam-combining spacecraft some six miles behind the ring. Instruments and equipment in yet another spacecraft, up to 30 miles behind the beam combiner, would measure the way those different x-ray beams interfere with each other and would create an image of the region just outside the black hole's event horizon. No one knows what such an image would show. But data from the Chandra X-ray Observatory, Hubble, and other current instruments paint a tantalizing picture.

"Probably we'll see some kind of flat disk and, more than likely, there are magnetic loops coming out of that disk very much like the loops we see on the surface of the Sun," believes Nicholas White, head of the High-Energy Astrophysics office at NASA's Goddard Space Flight Center. "These are releasing x-ray energy as material spirals in, big x-ray flares going off."

Images from the proposed array of 34 spacecraft—officially known as the Micro Arcsecond X-ray Imaging Mission, or MAXIM—also would likely show the formation of huge jets being blasted away from black holes like the one at the heart of galaxy M87.

"Somehow these things are forming jets that are shooting out at close to the speed of light from this very inner region," White says. "When we image it directly, we'll see that jet-formation region. We think it has to do with magnetic fields that are kind of being swept up by this disk, which is flying around the black hole and somehow funneling material out."

MAXIM is one of the most technically complex space missions ever conceived, requiring advances in control systems beyond anything available today. To perfect the technology, NASA plans to launch a quartet of satellites with x-ray spectrometers around 2010, in a mission known as Constellation-X, for which White is the project scientist. "That's kind of critical," he cautions. "We want to really make sure we understand the lay of the land before we get into the bigger stuff."

By about 2015, the MAXIM team wants to launch a two-spacecraft mission that would have 5,000 times the resolution of Chandra. Following this timetable, White says, a MAXIM-class x-ray interferometer with 10 million times Chandra's resolution could be put up by 2020 or 2025. The cost of such a project is unknown.

"With today's technology, obviously we're a long way from doing that," says White. "But we can get the engineers to start thinking about it. There are lots of elements of the technology that need to be developed. They know where we're trying to go."

While MAXIM promises to be a most unusual satellite array, it is still recognizably a telescope as well, for it will—like all telescopes—collect and concentrate some form of electromagnetic radiation. A second prospective tool for analyzing black holes, the $300-million Laser Interferometer Gravitational-Wave Observatory, or LIGO, is another beast altogether. It is more a hall of mirrors, designed to detect the way space itself may be stretched and compressed by gravity. This observatory consists of two stations, one in Louisiana and another in Washington state, each featuring a pair of 2.5-mile-long tunnels that meet at right angles. Test masses fitted with mirrors are suspended at both ends of each tunnel. In a building where the two arms meet, a beam splitter directs ultra-stable laser light into each tunnel, where it bounces back and forth between the mirrors. When the light is recombined, interference patterns result.

Physicists have pondered the nature of gravity for centuries. What is this strange attraction between masses? Many believe that gravity is a form of wave energy. If so, then gravitational waves emitted by truly catastrophic events such as collisions between distant neutron stars or black holes should be detctable on Earth. According to Einstein's general relativity theory, passage of a gravitational wave through LIGO would decrease very slightly the distance between the test masses in one tunnel and increase the distance in the other tunnel. The amount of this stretching and compression is small indeed: somewhere on the order of one thousandth of the diameter of a proton.

Still, that's enough to affect the interference patterns in the combined laser beams, giving scientists a way to directly "see" a gravitational wave for the first time. Because of the extremely small scale involved, two widely separated observatories were built to eliminate the possibility of a false reading.

LIGO is expected to begin scientific operations in the summer of 2002. Upgraded equipment will be installed in 2005 or 2006, increasing the sensitivity and extending LIGO's reach to events hundreds of millions of light-years away. Working with similar gravity-wave interferometers in Europe and elsewhere, astronomers eventually will be able to determine where, in general, any observed waves originate.

LIGO is an attempt to prove the reality of gravity waves, to determine their velocity and the nature of the subatomic particle that presumably carries gravitational energy. At the same time, it is also a powerful new tool for astronomers, one expected to provide insights into the processes at work in the most catastrophic events in the universe.

"It gives you a laboratory to study gravity that we've never had before," says Barry Barish, director of LIGO. "The second thing, likely to be truly groundbreaking in the long run, is gravitational astronomy. The large advances in astronomy in the last decade or so are because people have expanded the horizon and looked at multiple wavelengths. Looking at different messengers—neutrinos or gravitational waves—gives you yet more handles on the same objects."

As with conventional telescopes, the sensitivity of a gravitational wave detector is directly related to its size. Depending on the results of LIGO, NASA and ESA may launch a mission called the Laser Interferometer Space Antenna, or LISA, around 2008. Instead of beam paths 2.5 miles long, LISA would consist of three spacecraft arranged in an equilateral triangle measuring three million miles on a side.

Similarly, optical astronomy expects a big boost in the 21st century, as ground-based astronomers plan a new generation of monster telescopes that would have been inconceivable a few short years ago. To put these new scopes in context, the Next Generation Space Telescope, with a primary mirror 20 to 25 feet across, is the largest space-based optical telescope currently funded. Targeted for launch around the end of the decade, the NGST will remain operational for at least ten years and probably well into the third decade of the 21st century.

The world's biggest existing optical telescopes, the twin 10-meter (33-foot) segmented-mirror reflectors of the W. M. Keck Observatory in Hawaii, are enormously productive in their own right. Yet they are being further enhanced through optical interferometry, which creates even bigger virtual mirrors capable of extraordinary angular resolution. Still, virtual mirrors only go so far. For decades, astronomers have dreamed about building telescopes with multi-element segmented mirrors 80 feet or more in diameter. But making such giants has always been beyond the capability of technology—not to mention budgets.

That picture began changing in the 1990s, however, thanks primarily to advances in mirror casting and polishing techniques, computer-aided design and control systems and, most important, the advent of adaptive optics, which quickly and continuously counteracts the effects of atmospheric turbulence. By using devices called wavefront sensors to monitor the light from one or more guide stars, a computer can figure out how that light has been affected by its passage through different density layers in our atmosphere. Signals can then be sent to tiny electro-actuators in one or more deformable mirrors in the telescope's light path to exactly cancel out those atmospheric effects. The light is sampled tens to hundreds of times a second.

The bigger the telescope, the more complex are its adaptive optics, requiring more computer power and deformable mirrors with huge numbers of individually controlled actuators. But with the basic technology now in hand or at least within sight, U.S. astronomers hope to build and commission at least one and possibly two mammoth 100-foot segmented-mirror telescopes by 2012 or shortly thereafter. Six times wider than the fabled Hale Telescope on Mount Palomar and three times wider than either of the Keck twins, they would, with adaptive optics, have up to five times the resolving power of the Next Generation Space Telescope in some wavelength bands.

As if that's not enough, European researchers are studying the feasibility of making telescopes with truly gargantuan mirrors, up to 100 meters (328 feet) across. The biggest of these is called, appropriately, the Overwhelmingly Large Telescope, or OWL. Nothing like it

has ever been seriously contemplated before. With a segmented mirror longer than a football field, it would have 100 times the sharpness of Hubble and 10 times the combined light-collecting area of every optical telescope ever built! Cost? Probably $1 billion or more.

If it ever becomes reality, OWL's unsurpassed resolution and sensitivity would enable it to see details in galaxies that appear as mere smudges in the Hubble Deep Field. It could image the earliest possible supernova remnants and directly measure details about the expansion of the universe that are impossible with current ground-based instruments. OWL also could analyze spectroscopically the atmospheres of large planets orbiting nearby stars, to determine their chemical composition and detect signs of biological activity. In the words of Roberto Gilmozzi, team leader of the OWL project, creating such an instrument "would constitute a milestone comparable to that of the invention of the telescope itself."

The sheer scale of an OWL-class telescope is breathtaking. When pointing skyward, it would stand nearly 40 stories tall—roughly the height of the St. Louis Arch. Its mirror, made up of 1,000 to 2,000 hexagonal segments, would have a light-collecting area of nearly two acres. Funding such a monster likely would require international cooperation, and Gilmozzi estimates construction time of 10 to 15 years. But preliminary studies indicate no major technological breakthroughs would be needed, only development of faster computers to handle the adaptive optics system.

In case the 100-meter OWL proves unworkable, the Europeans are also considering somewhat smaller options. Astronomers at Sweden's Lund University, for example, envision building a 164-foot telescope. Its light-collecting area would be about half an acre—25 times larger than that of a single Keck mirror—and it would cost in the neighborhood of $630 million. According to the university's web site, "Adaptive optics will measure and correct the image quality at a large number of points. This will require one or more deformable mirrors with up to 100,000 actuators and advanced wavefront sensors to analyze image quality rapidly. In addition, to obtain a large field [of view], more than one deformable mirror is foreseen." The giant reflector would resemble a huge radio telescope, with its primary mirror (composed of 585 segments) taking the place of the dish. It would weigh some 5,000 tons and stand nearly 328 feet tall.

Meanwhile, the U.S.'s National Optical Astronomy Observatory is exploring potential designs for a 100-foot instrument known as the Giant Segmented Mirror Telescope (GSMT). This project ranks second only to the NGST in a National Research Council survey of national astronomy priorities through 2010. In its decadal report, the NRC's Astronomy and Astrophysics Survey Committee called it "a powerful complement to NGST in tracing the evolution of galaxies and the formation of stars and planets. The committee recommends that the technology development for the GSMT begin immediately and that construction start within the decade." The cost is not yet known, but is expected to be in the $500-million range.

Another 100-foot facility—the California Extremely Large Telescope, or CELT—is planned jointly by the University of California and the California Institute of Technology, which also operate Keck Observatory. Formal preliminary design studies are expected to begin in 2002, and CELT could become the first proposed superscope to actually become operable.

While its precise cost has not yet been determined, UC–Santa Cruz astronomer Terry Mast, who participated in CELT's conceptual design study, expects it to come in at "much less than half a billion dollars." He hastens to add, "I think we all have to acknowledge at this point that these are all gleams in the eyes of their creators and none have been built yet. The trick here for all of the projects is going to be making sure they are feasible from a cost point of view. To spend two billion dollars on a telescope at this stage of the game is probably not going to happen."

Steven Strom, of the National Optical Astronomy Observatory, notes that one of the major objectives of the super scopes will be to carry out what amounts to tomography on a cosmic scale. Astronomers dearly want to develop a three-dimensional picture of what the cosmos was like in its earliest stages; such images would help them determine how small-scale differences in the initial fireball of the Big Bang eventually gave rise to structures like galaxies.

Large telescopes, he explains, will be able to search the sky for the building blocks of galaxies and help determine "what they are, what they're made of, how massive they are—and then understand the basic processes by which stars form in these subunits and how stars form in the assembling larger units that ultimately compose galaxies like the Milky Way." This new generation of telescopes, he adds, also will shed light on just how solar systems evolve.

By 2015, at least one of the proposed superscopes should be operational; by the end of the third decade, an OWL-class instrument may be peering into deep space. A global network of ground- and possibly space-based gravitational interferometers will be working in concert to test relativity theory and to look for traces of radiation left from the birth of gravity itself, in the Big Bang. Optical- and radio-wavelength interferometers will continue to grow in size and sophistication. It is too soon to know whether a MAXIM-class x-ray interferometer mission will be up and running by the end of the third decade. But given the rapid advances in optical interferometry over the past ten years, it seems possible that one will get built eventually, giving astronomers the close-up views of black holes they need to decipher the physics of these cosmic enigmas.

Whether relativity theory and current views about the birth and evolution of the universe will withstand the scrutiny of these powerful instruments cannot be predicted, or course. But one thing is certain: Our knowledge about our place in the universe will grow by leaps and bounds as these new tools are brought to bear, revealing wonders that cannot even be imagined today. As with any era, old theories will fall and new ideas will take their place. But the search for answers will go on as astronomers puzzle out from afar the vast tapestry of the larger world around us.

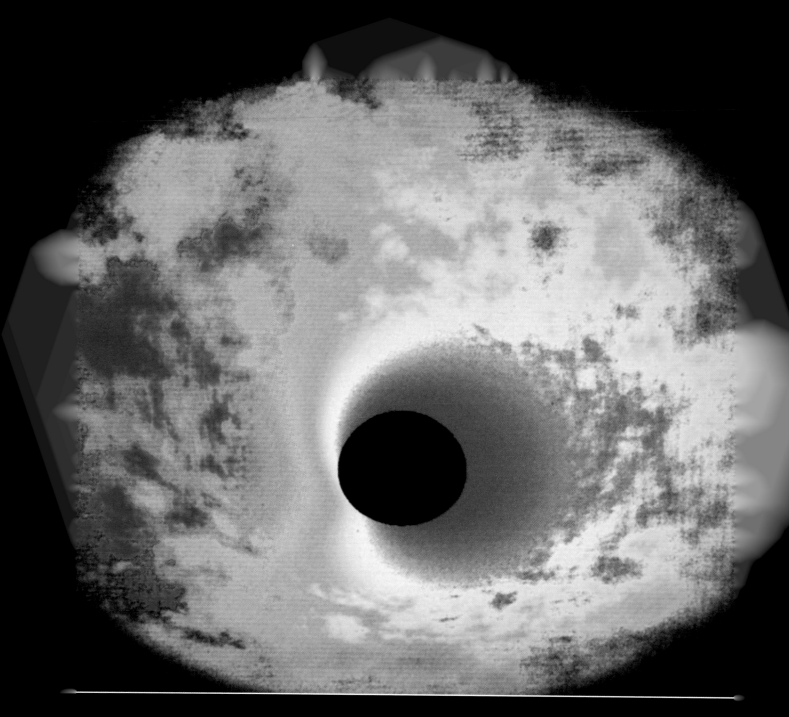

MAXIM—one of a new class of x-ray interferometer—promises to pack a million times the resolution of the Hubble telescope. While Hubble can detect large-scale jets shooting away from galactic black holes like M87 (right), MAXIM will be able to record radiation emitted near the invisible edge of the hole itself (above). It will utilize a fleet of 34 spacecraft, moving in precise formation and combining their data to work as one giant telescope. Theoretically, MAXIM and others like it will be able to resolve a feature the size of a dinner plate on the surface of the Sun.

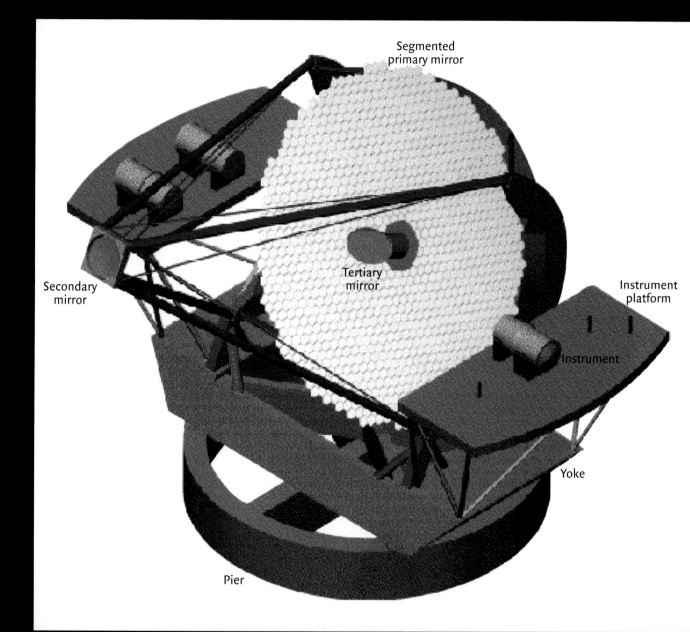

Segmented
primary mirror

Secondary
mirror

Tertiary
mirror

Instrument
platform

Instrument

Yoke

Pier

When built, CELT—the California Extremely Large Telescope—will be as big as a baseball field. At its core is a segmented primary mirror nearly 100 feet in diameter, made up of many small, hexagonal mirrors arranged in a honeycomb pattern. It will provide 10 times the light-gathering area of each of the main Keck telescopes, and will be used to probe nearby star-forming regions to study the birth of stars and perhaps detect extrasolar planets. At the other extreme, astronomers also will use CELT to examine some of the most distant objects, in effect peering far back in time to the early moments of the universe.

RESHAPING THE MOON AND MARS

By the end of the second decade of the 21st century, a spacesuited astronaut will step off a ramp to become the first human to visit another planet. That astronaut has already been born. The planet is Mars.

Given the current pace of planetary exploration, it's entirely possible that by the end of this century Martian colonists may be engaged in the greatest engineering project in human history: Turning the red planet's thin, frigid atmosphere into a warm, comfortably thick blanket of carbon dioxide-rich air. And a hundred years after that, descendants of the original colonists could be tending crops and enjoying afternoon strolls under a leafy canopy of trees imported from Earth.

"With today's technology, we could transform the climate on the planet Mars, making it suitable once more for life," Christopher

Lunar eclipse from the other side: Above the Moon's horizon, a distant Earth momentarily passes in front of the Sun, bathing the foreground in coppery tones that reveal rectangular solar arrays powering a lunar research station. Many astronomers and NASA scientists anticipate further exploration and eventual development of the Moon during the 21st century.

McKay, a NASA research scientist, writes in the March 1999 *Scientific American*. "Such an experiment would allow us to examine, on a grand scale, how biospheres grow and evolve. And it would give us the opportunity to spread and study life beyond Earth."

By mid-century, equally grand engineering projects could be underway on our Moon, with manned lunar bases constructing mammoth solar power stations or vast strip mines to extract helium-3 fuel for use in fusion reactors on Earth. While the colonization and eventual "terraforming" of Mars's atmosphere may not be sure bets this century, our voracious appetite for energy is. Humans currently consume about 14 trillion watts, or terawatts, of power. By 2050, a global population of 10 billion will demand 60 terawatts or more, according to the World Energy Council.

Fusion reactors offer a possible solution, using the same process that heats the Sun to liberate huge amounts of energy from relatively small amounts of fuel. There are challenges, however. For one thing, a self-sustaining fusion reactor—one that generates more energy than is needed to initiate the reaction— does not yet exist, despite a half century of effort. Perhaps 21st-century science and engineering, spurred by a looming global energy crisis, will find a way.

Another challenge is that the cleanest, most efficient fusion reaction involves the isotope helium-3. While only trace amounts of helium-3 exist on Earth, much larger quantities are available on the Moon, deposited over billions of years by the solar wind. According to University of Wisconsin physicist Gerald Kulcinski, all the electricity consumed by the United States in 1996 could have been provided by the amount of liquified helium-3 that can fit inside the cargo bay of a single space shuttle. Just one ton of helium-3 contains roughly three billion dollars' worth of energy. So it's worth going after. Extracting it is possible now, using current technology. But it would require engineering on a scale never before attempted in space. To obtain 2.2 pounds of helium-3, astronauts and automated machines would need to process the top 4 inches of 247 acres of lunar soil.

David Criswell, director of post-doctoral aerospace research programs at the University of Houston, believes he has a better idea: Build huge solar power plants on the Moon, beam the electrical energy they produce directly to Earth in the form of powerful microwave radar, and convert it back into electricity.

"If you want a prosperous world, it's the only option," he says. "The fundamental problem with commercial controlled nuclear fusion is it hasn't been demonstrated, except by the Sun, and second, you have to build a fusion plant. In contrast, the lunar solar power system is based on available, demonstrated technologies. It brings in net new energy that otherwise would just be radiated by the Moon into empty space."

The idea is simple. Build vast arrays of billboard-size reflectors that would serve as giant energy-broadcasting antennas. At least two huge lunar power stations would be needed to provide continuous service during the Moon's two-week, day-night cycle. During eclipses, supplemental ground-based generators, energy storage systems, or even fleets of reflector satellites could keep energy flowing until the lunar power stations once again "see" the Sun. At the proper wavelength, proponents say, the incoming radar beams would be no more than 20 percent of the intensity of noontime sunlight on Earth, and would not cause any significant environmental impact or health risk for people living near ground stations.

A similar plan, proposed decades earlier by Peter Glaser, suggested transmitting solar power not from the Moon but from satellites placed in Earth orbit. Then futurist Gerard O'Neill proposed building such satellites with lunar raw material, noting that "Any handful of lunar dust and rocks contains at least 20 percent silicon, 40 percent oxygen, and 10 percent metals. Lunar dust can be used directly as thermal, electrical and radiation shields, converted into glass, fiberglass and ceramics and processed chemically into its elements. Solar cells, electric wiring, some micro-circuitry and the reflector screens can be made out of lunar materials."

Criswell argues it makes more economic sense to go one step further and build the transmission antennas

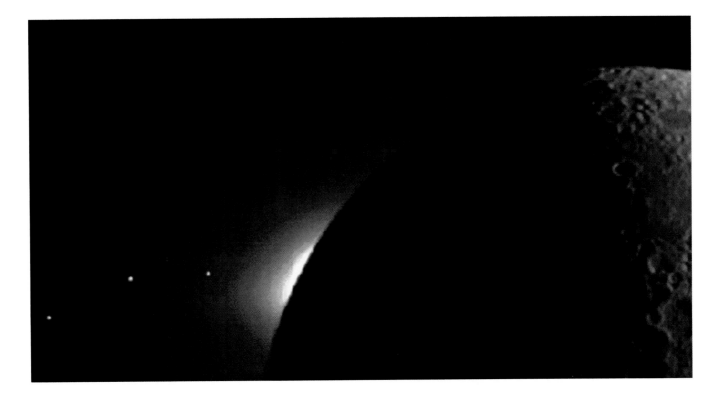

on the Moon itself. This system would provide more power for less cost per kilowatt, even though its transmitters and antennas would have to be much larger to compensate for the Moon's far greater distance.

Still, building such power stations would be an enormous undertaking, requiring a mature Earth-Moon infrastructure that includes large manned lunar bases and automated facilities. A demonstration plant that could generate up to 100 billion watts of power would cost around $20 billion and take a decade to build, Criswell estimates. A full-scale system generating 20 trillion watts of electrical power would cover thousands of square miles of lunar surface and cost trillions of dollars over the life of the program. But Criswell says it actually would be cheaper than any other current technology scaled up to that level of power. Regardless of the eventual size of such a system, building lunar power stations assumes the establishment of a permanent base on the Moon, which Criswell believes "is the next logical step for the world space programs after completion of the International Space Station."

Lunar scientist Alan Binder agrees. But he's not waiting for the government. Binder, the principal investigator for NASA's successful Lunar Prospector mission, wants to build a commercial Moon base that would rent out space to university scientists, government researchers, and anyone else who could afford it—and he wants to do it now. His idea is to launch a

Momentary lineup illustrates the ecliptic plane, the imaginary plane in which Earth orbits the Sun. Since our solar system's planets were formed from the spinning, flattened solar disk, they all tend to lie near the ecliptic. Saturn, Mars, and Mercury obediently circle the Sun at left, while Earthshine lights half of the lunar surface, at far right.

series of low-cost commercially financed reconnaissance missions to map the Moon and its resources in unprecedented detail. Data from such missions would be sold to NASA and others. A commercial base, he says, "would be the beginning of a lunar colony. But to do that, we have to understand the Moon and its resources, and gain experience."

Surprisingly, the Moon has not been mapped at the same level of detail as much more distant Mars. The locations of large features on the far side of the Moon, for example, are uncertain by six miles or more, according to Binder. He adds that engineers need to thoroughly assess the risk of Moonquakes and suggests that robotic landers deploy a network of seismic sensors, collect rock and soil samples from a variety of sites, and launch them to Earth for detailed analysis. Additional data would be gathered to determine how much water might exist in icy polar craters that never see the light of the Sun. Selection of a site with suitable

raw materials would follow, and a base would be built there capable of housing 12 to 16 people.

"We would simply rent out space to whoever wants to come—cosmonauts, astronauts, scientists, TV people," Binder says. "We could probably build a base for about $10 billion." He adds, "I firmly believe the Moon is the key to the rest of the solar system."

But for many, Mars, not the Moon, is the next logical step in exploring the solar system. Aerospace engineer Robert Zubrin has emerged as a leading advocate of manned Mars exploration; his "Mars Direct" approach, first proposed in 1990, turns traditional NASA mission planning on its ear. Instead of launching a manned, fully outfitted mission in a single bold—and expensive—stroke, he suggests slashing launch weights and costs by sending an expedition in stages and using the Martian environment to produce some of the raw materials needed for surface exploration and the trip back to Earth.

Using a heavy-lift Saturn 5-class rocket, he would first launch a 40-ton unmanned Earth return vehicle, or ERV, which would brake into orbit around Mars and then descend to the surface. Over the next ten months, a compact nuclear reactor on board would power an automated chemical plant that would use hydrogen carried from Earth and the Martian atmosphere to produce oxygen and methane fuel, both for the ERV and for future long-range ground vehicles.

Two years after the ERV's launch, two more Saturn 5-class boosters would go up, one with another ERV/fuel factory, the other carrying a habitation module with a crew of four. The crew would land at the site of the first ERV, now fully fueled, while the second return vehicle would land several hundred miles away and begin making the fuel needed by the next crew. "At the conclusion of their stay," Zubrin writes in his book *Entering Space*, "the crew returns to Earth in a direct flight from the Martian surface in the ERV. As the series of

The first astronaut to tread the Martian surface will likely find drifts and dunes of dust, according to data gathered by Mars Global Surveyor beginning in 1997. With its atmosphere, seasons, and raw materials, Mars is considered the logical planet for humankind to explore next—and perhaps even to colonize.

missions progresses, a string of small bases is left behind on the Martian surface, opening up broad stretches of territory to human cognizance."

Is NASA listening? You bet. But Douglas Cooke, manager of the advanced development office at NASA's Johnson Space Center in Houston, cautions that, for the first manned mission to Mars, "I can't imagine that we wouldn't send the crew without sending everything they needed with them. What you might do, though, is send a plant down that would make your fuel for the next mission. You'd have a crew there to intervene if there was a problem with the process and you'd know, with the crew there, that everything operated or it didn't and you'd have a much better feeling about the next time." He adds that NASA's initial manned mission to Mars likely will consist of a lander—launched unmanned from Earth—and a much larger mothership/return vehicle carrying a crew of five or six. The two vehicles would dock in Mars orbit and the crew would descend to the surface, where a pre-positioned habitat module would be waiting for them.

"If anything happened that caused the lander not to work or for some reason they couldn't dock, the crew would be able to come home," Cooke says. "They would have to stay in Mars orbit for a while, but at least they'd be in the vehicle they need to come home in."

Subsequent missions would use chemical processing plants on the Martian surface similar to Zubrin's, making fuel needed by the lander to return to Mars orbit and link up with the mothership for the flight home. Cooke comments on Zubrin's plan, "He manufactures all the fuel to get all the way back to Earth. To do that, he ends up sending the crew back in a very small vehicle. After a very long mission to begin with, to confine them to very cramped quarters to send them all the way back home, we wouldn't do it, I don't think."

Both plans call on astronauts to spend a year and a half or so on the Martian surface. It's not known, of course, when such a mission might be launched. But Cooke says even NASA's more conservative, more expensive scenario could be implemented by the middle of the second decade—if government gives the go-ahead. Beyond that, much will depend on what the first Mars astronauts find as they explore their strange new world. The discovery of life there or of fossils of extinct life forms would have a profound impact on thought and likely would spark more exploration. But even without such discoveries, Mars remains a logical step in humanity's exploration of the solar system. It is the only world besides Earth that offers a remotely similar environment with a 24-hour day/night cycle, an atmosphere, seasons, and the raw materials necessary to support a technological civilization.

And it is the only planet that can be terraformed with today's technology to make it more hospitable to Earth-evolved life. Christopher McKay says terraforming could be accomplished by building small robot factories on Mars that would pump heat-trapping "super greenhouse" gases such as perfluorocarbons, or PFCs, into the Martian atmosphere. Such gases are thousands of times more efficient at trapping solar energy than carbon dioxide, the most abundant greenhouse gas on Earth or Mars. Calculations show that tiny amounts of super greenhouse gases—a few parts per million—would be enough to raise the average

temperature of Mars from -76°F to 40°F. Such warming would release more carbon dioxide, currently locked up in the Martian soil and polar ice caps, which would accelerate the process and lead to the release of even more CO_2 in a sort of positive feedback loop.

"To generate enough greenhouse gases, we would need to distribute hundreds of small PFC factories across the Martian surface," McKay writes in *Scientific American*. "Powered by solar energy, each of these Volkswagen-size machines would harvest the desired elements from Martian soil, generate PFCs and pump these gases into the atmosphere."

Within 100 years, he adds, such terraforming "could generate a thick carbon dioxide atmosphere... and lead to a water-rich planet in about 600 years." Humans could not breathe such air, but could forego bulky spacesuits in favor of simple oxygen masks. Plants and microorganisms would thrive.

"On Mars, the bare rock would give way to the hardy plants that thrive on Earth's tundra, and eventually the Martian landscape would blossom into the equivalent of an alpine meadow or a pine forest," McKay continues. "The plants would generate oxygen, and eventually insects, worms and other simple animals that can tolerate high concentrations of carbon dioxide and low levels of oxygen could roam the planet." Given enough time—say a million years—photosynthesis by imported plants from Earth could pump enough oxygen into the rejuvenated atmosphere to make it breathable by humans, a process that took more than two billion years on our home planet.

This scenario assumes future astronauts do not find any native organisms living on Mars. If they do, terraforming could be tailored "to allow that native life to emerge and spread across the planet," McKay writes. "But on Mars, in all likelihood, no life forms currently exist. Thus, any biological expansion would be considered an improvement. If spreading life is the objective, making Mars habitable might allow humans to make a purely positive contribution for once." It is a grand vision. A chance to build a new world. And it is within the grasp of children who are alive today.

Robots lead the way in exploring Mars: A lander (right) analyzes soil samples and beams results back to Earth, while a mobile rover fetches more distant samples. Another lander, having finished its search, blasts off (above) to begin its return to Earth. For manned missions to Mars, NASA envisions first landing an automated, fuel-generating habitat on the planet, then sending a crew in a mothership with its own lander to ferry them to the surface. Some even consider terraforming the red planet—modifying it to make it fit for humans. Within a hundred years, they say, small robot factories on Mars could create a thick carbon dioxide atmosphere that would raise the planet's temperature and allow plants and microorganisms to thrive. Colonists could live and work wearing only simple oxygen masks.

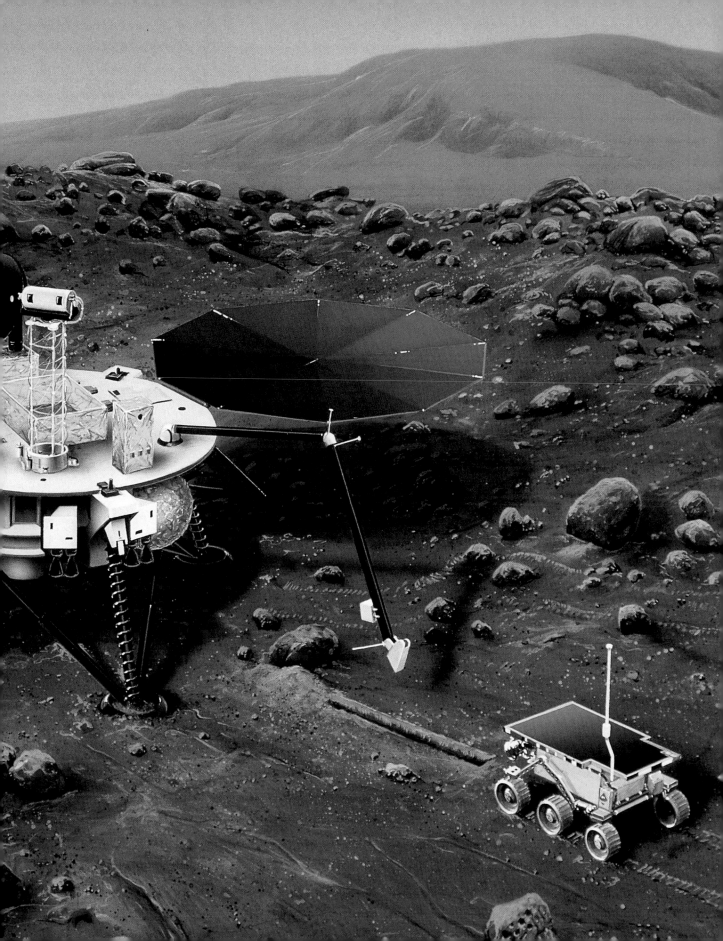

SETI, INTERSTELLAR TRAVEL, AND BEYOND

Before the 21st century ends, humans will know whether life ever evolved on Mars, under Europa's icy crust, or anywhere else in the solar system. We will know how common Earth-like planets are in the Milky Way, what percentage possess biospheres, and, through spectroscopic analysis of their atmospheres, whether any planets in relatively nearby solar systems harbor civilizations even remotely like ours. If so, and if any alien cultures near or far broadcast electromagnetic signals, either deliberately or inadvertently, the ongoing Search for Extraterrestrial Intelligence— SETI—will have found one or more of them and perhaps initiated replies in the opening rounds of a multi-generational dialogue.

Closer to home, we will have thoroughly explored our own solar system, establishing robotic mines and research stations, as well as

"Where are they?" asked physicist Enrico Fermi of extraterrestrials 50 years ago, posing one of mankind's most intriguing questions. Many believe SETI—the Search for Extraterrestrial Intelligence— may finally achieve its goal during this century. Here a 70-meter (229.6-foot) radio antenna turns its powerful ear toward a distant star's glimmer.

commercial bases—if not self-sufficient colonies—on the Moon and Mars, in the asteroid belt, and beyond. High-efficiency ion drives, sailcraft driven by sunlight or laser beams, and perhaps fusion-powered rockets and even antimatter drives likely will power the first robotic interstellar probes to nearby stars, surely one of the crowning technological achievements of this century. Such systems will provide the transportation backbone of a truly spacefaring civilization.

In *The Starflight Handbook*, authors Eugene Mallove and Gregory Matloff warn, "The vast distances that separate the stars loom as a barrier.... Even so, precursor missions to explore near interstellar space will probably happen by the mid-21st century, if not sooner. Much faster probes and later missions bearing people will be launched toward the stars when advanced propulsion systems now only theoretically possible come to fruition."

Such projections may read like science fiction but they are reasonable, based on past scientific and technological progress as well as the promise of emerging technologies already in limited use—assuming, of course, that humanity does not destroy itself along the way or suffer global economic or environmental disasters that might quench the drive to explore.

And there's a wild card in this heady scenario. It's impossible to know what theoretical breakthroughs might occur over the next 100 years. While faster-than-light travel appears impossible in the context of Einsteinian relativity, one cannot rule out the possibility that someone eventually will figure how to move an object from point A to point B faster than a speeding photon, opening the galaxy to human exploration. And if this doesn't occur, exploration and exploitation of our solar system can be achieved within the context of known physics. Though scientific challenges abound, no major theoretical breakthroughs are required, such as relativity theory and quantum mechanics, both of which emerged at the dawn of the 20th century.

But exploring the solar system is child's play compared to setting sail across the incomprehensible ocean of interstellar space. For that to occur, major

theoretical breakthroughs may, in fact, be necessary to turn the dream of starflight into reality.

Consider again the sheer size of the universe. Imagine that one inch represents the distance from the Earth to the Sun—93 million miles, or one astronomical unit. At that scale, one mile equals one light-year. Pluto lies a little more than a yard from the Sun; our entire solar system is about as wide as a professional basketball player is tall. Even on this scale—our solar system reduced to the size of a human being—the nearest star, Proxima Centauri, lies 4.3 miles distant. And our home galaxy sprawls 100,000 miles across.

Now consider NASA's Voyager 1 probe, currently departing the solar system at about 3.6 astronomical units per year, or 38,590 mph relative to the Sun. If it were headed toward Proxima Centauri, it would need some 75,000 years to get there.

"If such a probe had been launched from Earth the day *Homo sapiens* first set foot in Europe, it would still have another 30,000 years to go," writes engineer and Mars visionary Robert Zubrin in *Entering Space*. "In the face of such imposing challenges, a literature has been created showing interstellar travel as dependent on the exploitation of exotic and fantastical physical phenomena such as wormholes, space warps, cosmic strings and so forth. While some of those concepts are mathematically consistent with the currently known laws of the universe, there is no evidence that they actually exist or, if they do, that there is any method by which they could be manipulated by humans to produce a practical technology for space propulsion. Therefore, many people believe that interstellar travel is impossible."

Zubrin and others disagree, however, arguing that interstellar travel can be accomplished within the framework of known physics even though it may prove as difficult "as a flight to Mars would have appeared to Christopher Columbus."

Chemical rockets like those in use today are out of the question, as are spacecraft based on nuclear fission rather than fusion. While fission-powered rockets could theoretically carry robots or people to nearby stars, such voyages would take centuries to complete because of the relatively low efficiency of such systems. But even that does not rule out star travel. A frequent theme in science fiction is the "arkship," a huge self-contained flying city that carries generations of star voyagers across the depths of space. The concept has many variants, including smaller robotic "slowboats" carrying frozen embryos that could be thawed and reared to adulthood on a suitable planet decades or even centuries later, or spacecraft staffed by astronauts who would spend most of the journey in some form of hibernation or suspended animation.

Actually, the propulsion technology for slowboats may be easier to achieve than the rest. Research into artificially induced hibernation or suspended animation is in its infancy, life support systems capable of operating unattended for centuries do not exist, and the hardships imposed by such severe cultural isolation have not been explored. Fusion drives, however, could shorten trip times to a single human life span, at least in theory. Still a tall order, but one inherently more palatable than voyages lasting generations.

Again, several innovative variants have been proposed. One could detonate small fusion bombs behind a spacecraft to accelerate it to enormous velocities. Or lasers could be used to ignite fusion fuel pellets, accomplishing the same goal. Alternatively, Zubrin writes, one could direct super-heated plasma out one end of a fusion reactor to achieve theoretical velocities up to 10 percent that of light. Such a spacecraft could accelerate and then brake into Proxima Centauri's solar system—the one nearest ours—in less than 100 years.

Light sails also might work. In 1962, the physicist and writer Robert L. Forward proposed pushing a spacecraft to extreme velocities using the concentrated light of a gigantic laser. A 1,000-ton spacecraft with an ultrathin sail 430 miles across could be boosted to 15 percent light speed in just two months by a laser five times as bright as sunlight on Earth. That's fast enough to reach Proxima Centauri in only 29 years.

But building and operating that laser-sail system presents enormous challenges. The materials for such a sail do not exist, nor does the laser. And if they did,

some 240 billion watts of power would be required to fire the system, Zubrin writes—"about 20 times the total power humanity currently generates each year." Still, if humanity's power production continues to increase at current rates, such an expenditure might not be out of the question in another century or two.

Rockets powered by antimatter offer perhaps the ultimate boost in the known-physics category, 1,000 times more energy than fission reactions and 100 times more than fusion. Antimatter consists of negatively charged protons and positively charged electrons, or positrons, the opposite of normal matter. When matter and antimatter meet, they annihilate each other in a flash of energy. Lots of it. The mutual annihilation of a single pound of matter and antimatter would release as much energy as 18 million tons of TNT. Flights at up to 90 percent the speed of light "would get a ship to Alpha Centauri in about five years, which would seem like three to the crew due to the effects of relativistic time dilation," Zubrin writes. "But the society that launched such a mission would have to be one that was so rich that cost was simply not an issue."

That's because the cost of producing antimatter is truly extreme, at least now. Mallove and Matloff figure it takes roughly $100 billion to produce a single milligram of antiprotons, only 0.000035 ounces. Forward suggests it must fall to $10 million per milligram—one ten-thousandth of the current price—if antimatter propulsion is to become economically viable. That assumes, of course, the technological hurdles can be overcome and a spacecraft can fly unscathed through light-years of interstellar dust and debris.

"You have a lot of particles out there and they're going to be hitting you at relativistic speeds," warns George Schmidt of NASA's Marshall Space Flight Center. "Some studies have shown you're going to completely obliterate your craft long before you get up to those speeds. Other studies are saying it's perhaps not that much of an impact. No one really knows."

Many concepts have been developed that actually take advantage of the interstellar medium. Robert Bussard, working at Los Alamos in 1960, came up with

the idea of an interstellar fusion ramjet. Now known in science fiction circles as the "Bussard ramjet," it uses a huge magnetic funnel to scoop up electrically charged interstellar particles and direct them into a fusion reactor, where they are burned to produce thrust.

With a virtually unlimited fuel supply, a Bussard ramjet could accelerate continuously, pushing ever closer to light speed, and time would slow to a crawl. Astronomer Carl Sagan once calculated that a fusion ramjet could reach the Pleiades star cluster, 400 light-years away, in just 11 years of shipboard time. But the interstellar ramjet may not be possible using the fusion reactions proposed by Bussard. Certainly such technology does not appear to be on this century's horizon.

In the end it may be a moot point. Looking for a shortcut around Einstein's speed limit, a small but growing number of researchers is poking into obscure corners of relativity in search of theoretically permissible ways to circumvent the light-speed restriction that appears to be built into the fabric of the universe. At the Glenn Research Center in Cleveland, Marc Millis oversees the Breakthrough Propulsion Physics Project, established in 1996 to explore and further theoretical concepts for advanced propulsion technologies.

"Even if the pessimistic story is the case and all of these things turn out to be flatly impossible, that the only way we can propel spaceships is with rockets or light sails or whatever, by trying to find these alternatives we're going to create more useful information than if we just give up," Millis says.

So-called wormholes offer one such tantalizing prospect, allowing travelers utilizing quantum short-circuits in the space-time continuum to tunnel from one point to another in no time at all—or even less than that. Imagine a flat sheet of paper with two points labeled A and B. A conventional traveller would start at A and move two-dimensionally across the paper to B, taking some finite amount of time. Now fold the paper

Colonists ultimately may thrive on other worlds, but getting them there remains an awesome challenge. Technological breakthroughs will be necessary for propulsion systems to reach even the closest stars. Here, pioneers have settled a nameless moon with self-contained outposts that presumably feature all the comforts of their home planet.

so that A and B are touching. Our intrepid wormhole traveller would no longer need to physically move across the paper. The origin and destination would be contiguous; with one step, our traveller could move light-years. The trick, of course, is figuring out how to fold the paper—how to create such a wormhole in space-time, assuming such things even exist. And maybe they do. In 1988, physicists at the California Institute of Technology carried out calculations indicating wormhole transportation is at least possible in principle. Even more intriguing, their existence raises the possibility of moving backward in time!

"The researchers suggest that a space-time wormhole, the byproduct of black hole formation, could be maintained by employing specially placed charged spheres," write Mallove and Matloff. "One way of creating such a wormhole, they say, is to pluck out and amplify to macroscopic size a microscopic fluctuation in the universe's presumed underlying quantum mechanical 'space-time foam.'" Sounds easy, right? But maybe not in the next 100 years.

Such exotic concepts are in keeping with the three stated goals of the Breakthrough Propulsion Physics Project: To discover methods that eliminate or sharply reduce the need for propellants, to discover if it might be possible to achieve or exceed the speed of light, and to find new sources of energy to power starships.

"My personal feeling, and I try not to let this taint my work, is I'm pretty pessimistic on number two," says Marc Millis of faster-than-light travel. "The first one, finding some other way to create forces, there are enough unexplored, even old historic physics stuff where there are perhaps back doors to that issue and a few anomalous effects which still haven't really been sorted out where that might be achievable."

In the end, practical interstellar travel may not be possible. Or if it is, it may be so incredibly difficult and expensive that it is simply not an option for direct exploration, restricting humans and alien races to the limited resources of their own solar systems. In fact, one could argue that if such travel were possible, the galaxy would be teeming with interstellar civilizations and Earth almost certainly would have been visited many times in the past. Contrary to the claims of UFO enthusiasts, there is no conclusive evidence that Earth has ever been visited by aliens.

The same logic can be applied to the search for extraterrestrial intelligence in general. In a population of 200 billion stars, if only one solar system in a million developed intelligent life over the course of the Milky Way's history, some 200,000 civilizations would have arisen by now. But there is no direct proof that more than one—ours—has ever evolved. The absence of radio signals and other clear evidence of past or present alien encounters are what prompted physicist Enrico Fermi to pose a simple question more than 50 years ago: "Where are they?"

What has become known as Fermi's Paradox has many possible answers. But, as astrophysicist Jeffrey Bennett writes in *On the Cosmic Horizon*, they fall into three basic categories:

1. Galactic civilizations do not yet exist. Civilizations in general are not common and many planets remain unexplored;

2. Interstellar travel is not possible, or races tend to destroy themselves before achieving it;

3. Galactic civilizations exist but have deliberately concealed their presence from Earth, either because humans are considered potentially dangerous or too immature to join the galactic fraternity.

Barring direct alien intervention or detection of a signal, however, it is not possible to determine which, if any, scenario might actually be correct. But scientists currently carrying on the 40-year-old Search for Extraterrestrial Intelligence—SETI—may soon be able to answer one of the most profound questions in modern science: Are we alone in the universe?

"The search has just begun," says Douglas Vakoch, a psychologist at the SETI Institute in Mountain View, California. "Even if primitive life originates on another planet, what are the series of steps necessary for it to become intelligent and not only have the technology but be motivated to use it? So I think given the number of stars that we have looked at so far, it would be premature to say, 'Wow, no one's transmitting.'"

The idea of attempting to remotely detect alien civilizations by intercepting deliberate or inadvertent radio transmissions was proposed in 1959 by Cornell University physicists Giuseppe Cocconi and Phillip Morrison. The following year, radio astronomer Frank Drake used an antenna at Green Bank, West Virginia, to listen for radio emissions from two nearby stars, Tau Ceti and Epsilon Eridani. Project Ozma, as it was known, did not detect any intelligible signals. Dozens of other searches have been carried out since then, all with negative results. But they have barely sampled the vast number of possible frequencies a civilization could be using. Researchers are hopeful that advances in computer processing power and the use of ever larger antenna arrays eventually will enable them to find one or more needles in the cosmic haystack.

"In the long run, I think it's inevitable a detection will occur, unless there is some completely unbelievable reason that life here on Earth is unique," says Tom Pierson, executive director of the SETI Institute. "We take a very long-range view of it. The most commonly used phrase is this is a multi-generational search. It may be hundreds of years. And even if there's a signal detected 10 years from now or 50 years from now or one day from now, the search will continue. In fact, the search might intensify."

The SETI Institute's Phoenix Project is the most powerful search currently underway, utilizing radio

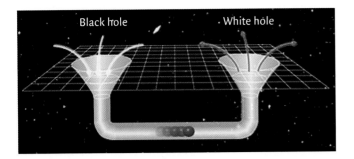

Black hole **White hole**

Wormholes offer a highly attractive but also highly speculative shortcut through space. In theory, material sucked into black holes may be expelled out of other structures called white holes; hypothetical space-time tunnels connect the two. In a spaceship negotiating a wormhole at something near the speed of light, time might travel backwards.

telescopes in both hemispheres to search 1,000 nearby stars. The equipment is sensitive enough to detect the equivalent of a strong terrestrial radar signal from 150 or so light-years away. Untargeted sky surveys also are underway, piggy-backed on other radio astronomy projects, along with modest searches for possible optical or infrared communications signals.

Even the public is actively involved. The SETI@home project automatically distributes chunks of raw data collected by the Arecibo radio telescope in Puerto Rico to the personal computers of more than three million volunteers in more than 200 countries. When the owner is not using his computer, the SETI@home software uses his computer to analyze the Arecibo data for patterns, working in the background as a high-tech screen saver. Results are sent through the Internet to a central computer and another data chunk is delivered.

Long-range goals for the SETI Institute include developing a large, dedicated antenna array for targeted searches, developing targeted optical and infrared searches, and developing a large antenna array capable of detecting strong signals from anywhere in the sky. In August 2000, it received a $12.5 million boost from Microsoft co-founder Paul Allen and Nathan Myhrvold, Microsoft's former chief technology officer. The Allen Telescope Array will be located at Berkeley University's Hat Creek Observatory, in northern California. Made up of hundreds of 20-foot-wide dish antennas, it will have a collecting area of nearly 2.5 acres. It is expected to become oper-

ational in 2005. Using this powerful new tool, SETI researchers will be able to carry out a targeted search of 100,000 stars in just six years.

The SETI Institute also is participating in international studies of a proposed monster telescope called the Square Kilometer Array, or SKA, that would have a collecting area of 247 acres. Such a multi-antenna radio interferometer, expected to cost up to $1 billion, could search a million or more stars for possible signals from intelligent aliens. Radio astronomers hope to have a SKA-class instrument in operation by 2015 or 2020. Says Vakoch, "There are a lot of steps necessary to get an intelligent civilization that can and wants to make contact. But...I am struck by the diversity of life just on this one planet. So I remain optimistic. Even if there's a one-in-a-thousand shot, a one-in-a-million shot, I am very happy to be involved in this project. Because I think the payoffs would be tremendous."

Possible contact with alien beings. Colonization of the solar system. The first slow steps away from the cradle of Earth into the great gulf of interstellar space. Barring economic collapse or environmental disaster, humankind has a more-than-reasonable chance of turning today's science fiction into 21st century fact.

Of course, history is littered with predictions that went too far and expectations that were out of step with reality. Air cars come to mind, even Stanley Kubrick's beautiful rotating space station in 2001: A Space Odyssey. Who would have imagined in 1968, when that movie premiered at the height of the Apollo program, that humans would still be limited to low-Earth orbit when 2001 finally rolled around? "It's sort of like everybody went to sleep for 30 years," says NASA's Christopher McKay.

But the Moon program was the product of politics, a Cold War space race that could not be sustained. Our future space explorations will be driven by economic and environmental imperatives, perhaps by the need to exploit distant resources essential to an ever expanding population with an insatiable appetite for energy. In all likelihood, we won't have any choice. And who knows? We might even meet an alien or two along the way.

In this futuristic scene, artist Gary Tonge combines imagination with 21st-century computer technology to convey the possibilities and what he calls "the indescribable fantastic magnitude" of the universe. An advanced civilization has built mushroomlike spaceship garages that extend far beyond the atmosphere of a recently colonized planet. Overhead, distant galaxies take shape in a nebula. Such worlds may be waiting just around the next supercluster as man embarks on his greatest odyssey.

absolute zero A temperature of minus 273°C or zero kelvin; the point at which all molecular activity ceases.

antimatter Material made of atomic particles that are the precise opposites of those in ordinary matter.

antipode Any point diametrically opposite from another point on a planet.

antiproton A form of antimatter; the antiparticle of a proton.

aphelion The point in a solar orbit farthest from the Sun; the opposite of perihelion.

apogee The point in a satellite's orbit which is farthest from the mother body; the opposite of perigee.

apparent magnitude The brightness of a star as seen from Earth.

apparent motion Movement of a celestial body in relation to more distant ones.

asteroid A small rocky object which orbits the Sun, usually in the asteroid belt between Mars and Jupiter.

astronomical unit (AU) A unit of length equal to the mean distance between Earth and the Sun, approximately 93 million miles.

atom The smallest unit of a chemical element, consisting of a nucleus (protons and neutrons) and orbiting electrons.

aurora A glow in a planet's upper atmosphere, caused by the interaction of the solar wind, the planet's magnetic field, and atoms in the atmosphere.

Big Bang The theoretical origin of the universe some 18 billion years ago, when all matter and radiation was thrust outward from a single point.

binary star One of a pair of stars orbiting the same center of gravity.

black dwarf A theoretical state for a white dwarf that has radiated away all its energy and is only detectable by its gravitational pull.

black hole A massive collapsed star whose gravity is so strong that neither matter nor radiation can escape it.

brown dwarf A "failed star;" similar in composition to a star but too small to become hot enough to fuse hydrogen into helium.

caldera A crater formed by volcanic activity.

Cepheid variable A type of very bright star that pulses in a highly regular pattern and thus can be used to measure its distance from Earth.

closed universe A universe that eventually reverses its expansion, collapsing on itself.

comet A relatively small luminous mass of ice, dust, and solid matter orbiting the Sun.

constellation One of 88 arbitrary groupings of stars that human cultures identify with a person, animal, or object.

corona The Sun's hot, highly ionized, luminous outer atmosphere.

coronal mass ejection (CME) A huge expulsion of plasma from the Sun, often associated with solar flares.

cosmology The study of the origin, evolution, and structure of the universe.

dark cloud nebula A type of nebula which absorbs light and hence is dark, such as the Horsehead nebula.

dark matter Galactic matter that cannot be seen, because it neither emits nor reflects light; the bulk of the known universe consists of this strange type of matter.

Doppler effect A shift in wavelength observed when the source and the observer are in relative motion.

eclipse Total or partial blocking of light by one celestial body passing in front of another.

emission nebula A cloud of gas unassociated with a particular star but hot enough to give off light.

escape velocity The speed at which one object must travel to escape the gravitational pull of another.

fission The splitting of an atomic nucleus into constituent particles, which generates energy.

flat universe A universe in which expansion slows, coming to a stop over an infinite amount of time.

fusion The combining of small atoms to create larger ones, with the production of energy.

galaxy A massive, disk-like assembly of stars, dust, and gas, held together by mutual gravitation.

Galilean moons The four largest satellites of Jupiter: Io, Europa, Ganymede, and Callisto, discovered by Galileo Galilei (1564-1642).

gas giants Four of the outer five planets of the solar system: Jupiter, Saturn, Uranus, and Neptune.

giant A star roughly 100 times as large and bright as the Sun.

globular cluster A spherical grouping of many thousands of mature stars within a galaxy.

gravitational lensing The bending of light from a distant galaxy or quasar by the gravitational pull of a nearer galaxy or cluster.

gravity The attraction of a celestial body for all other bodies, due to its mass.

heliopause The outer boundary of the solar system, roughly 100 AU, where the solar wind and the Sun's magnetic field join the magnetic field of the galaxy.

Hubble's Law The tenet that the farther away a galaxy is, the faster it is moving away from us.

intrinsic brightness the brightness of an object independent of its distance.

ion An atom that has an electrical charge because it has lost or gained one or more electrons.

Kuiper belt A band of widely scattered icy chunks that orbit the Sun beyond Neptune but inside the Oort cloud.

Lagrangian point A location in space where the Sun's and Earth's gravitational forces perfectly offset each other.

light-year The distance light travels in one year (about 6 trillion miles, or more than 60,000 AU).

MACHO Acronym for massive compact halo object; a component of dark matter wind that produces gravitational lensing.

Magellanic Clouds (Large and Small) The two irregular galaxies nearest the Milky Way, both easily visible from the Southern Hemisphere.

magnetosphere The region of space in which a planet's magnetic field dominates the solar wind.

main sequence star A star type that falls in the middle of the Herzsprung-Russell diagram, neither a giant nor a dwarf. Includes our Sun and most stars.

meteor An object that burns upon entering Earth's atmosphere, glowing brightly enough to be called a fireball or "shooting star."

meteorite The part of a meteor that survives the incendiary trip through Earth's atmosphere and lands on its surface.

meteoroid A small body that orbits the Sun.

Milky Way Our home galaxy, perhaps 100,000 light-years across, containing hundreds of billions of stars; also refers to the concentrated band of stars visible across the night sky.

nebula Latin for cloud; a cloud of interstellar gas and dust. Originally applied to any celestial body (apart from comets) that could not be resolved into a point of light.

neutron star Final stage in the life of a star no greater than three solar masses; it is extremely dense and is believed to result from a supernova. Smaller stars become white dwarfs; larger ones become black holes.

Oort cloud A spherical region of the solar system, beyond the planets and Kuiper belt, roughly 40,000 to 50,000 AU from the Sun; a vast reservoir of comets.

open universe A universe that continues to expand indefinitely because the gravity of its components is insufficient to slow the initial force of the Big Bang.

orbit The path a satellite takes, due to gravity or other forces, around a larger, more massive body.

orbital period The time it takes an object to complete its orbit.

perigee The point in a satellite's orbit that is nearest to its mother body.

photosphere The visible surface of the Sun.

planet An astronomical body larger than an asteroid but smaller than a star, and thus incapable of supporting self-sustaining fusion.

planetary nebula A type of nebula characterized by expanding and colliding gas clouds around a single hot star; a stage in the life of a main sequence star as it evolves from a red giant into a white dwarf.

planetesimal One of countless bodies of primordial gas and dust from which the planets and asteroids aggregated.

plasma The fourth state of matter, consisting of hot and highly ionized gas.

proplyd Protoplanetary disk; a disk of material often observed around young, formative stars that is believed to give rise to planets of that star

pulsar A rapidly rotating neutron star, which emits pulses of radio waves and light.

quasar A very distant starlike object with a large redshift; short for quasi-stellar object.

radio galaxy The most common and least energetic type of active galaxy, which emits large amounts of radio waves but little of other wavelengths.

red dwarf A star less massive and bright than the Sun.

red giant A relatively cool star that has ceased fusing hydrogen into helium and swelled to many times its original size due to reduced gravitational attraction.

redshift A shifting of spectral lines toward the red end, caused by the visual equivalent of the Doppler effect due to expansion of the universe.

revolution The motion of one body orbiting another.

rotation The turning of a body on its axis.

SETI Acronym for Search for Extraterrestrial Intelligence; the hunt for electromagnetic signals that could indicate other forms of life in the cosmos.

solar flare A violent release of magnetic energy, often accompanied by large amounts of radiation and charged particles; associated with sunspots.

solar nebula The whirling disk of gas and dust from which the solar system is thought to have taken shape.

solar prominence A glowing mass of solar material erupting from the Sun's surface and shaped by the solar magnetic field.

standard candle A celestial object whose intrinsic brightness is known, thus its distance can be calculated and used to estimate distances to its galaxy. Examples are Cepheid variables and Type Ia supernovae.

supernova A violent stage in the life of a star that is more than nine times as massive as our Sun; characterized by a huge release of electromagnetic radiation. A Type Ia supernova results from the explosion of a white dwarf; a Type II results from the collapse of a massive star.

terrestrial planets The four planets orbiting nearest the Sun: Mercury, Venus, Earth, and Mars.

white dwarf The final stage in the life of a main-sequence star that has exhausted its fuel and shines from residual heat.

WIMP Acronym for Weakly Interacting Massive Particle; a subatomic particle that travels freely through matter and may be a component of the dark matter wind.

x-ray star A type of star that emits radiation in the x-ray range (shorter than ultraviolet, longer than gamma rays).

William Harwood grew up in Nashville, Tenn. He holds a journalism degree from the University of Tennessee and joined United Press International in 1982. He was named UPI's Cape Canaveral, Fla., bureau chief in 1984, covering space shuttle operations, unmanned civilian and military space missions, as well as deep space exploration. He began working for CBS News in 1992, helping the network coordinate its coverage of space stories and serving as an on-air consultant. He also covers the shuttle program for the *Washington Post* and *Astronomy Now* magazine. He lives in Florida with his wife Catherine, a free-lance writer and former space reporter. They have two children.

ACKNOWLEDGMENTS

The author and the Book Division wish to thank the individuals, groups, and organizations mentioned in this publication for their guidance and assistance. This book would have been impossible without the help and support of Al Ortiz and Mark Kramer at CBS News Special Events; Rob Navias at the Johnson Space Center; Mary Beth Murrill, Guy Webster, and Jane Platt at NASA's Jet Propulsion Laboratory; Donald Savage and Dolores Beasley at NASA headquarters; Doug Isabell at the National Optical Astronomy Observatory; and Nancy Neal at the Goddard Space Flight Center. Special thanks to Reta Beebe at New Mexico State University; David Leckrone, Hubble Space Telescope senior project scientist; Rachel Akeson at JPL; Evan Smith at Goddard; and my editor, Tom Melham, for his vast patience. Special thanks also go to Van and Corinne Bunch for use of the Big South Fork facility, to Hal Bibee and the gang at Wilderness Resorts, and to Lyn Clement. Extra special thanks to my wife Catherine, for reading over my shoulder and cheerfully taking care of business during this project, and to Houston and Riley. Finally, thanks to my parents Kate and Walter Harwood, for encouraging their children to wonder about the universe.

Cover: Photo Researchers/Chris Butler/Science Photo Library; 2-3 ©Anglo-Australian Observatory/Royal Observatory, Edinburgh; 4-5 Wally Pacholka/AstroPics.com; 7 Photograph by Tom Sebring/©HET Partnership; 8-9 Don Foley (Milky Way illustration by Kenneth Eward/BioGrafx); 10 Naoyuki Kurita; 12-13 Doug Duncan; 14 Image by Craig Mayhew and Robert Simmon, NASA/GSFC; 17 NASA; 18 (top) NASA; (bottom, both) ©Calvin J. Hamilton; 19 (top) NASA; (bottom, left) NASA/JPL; (bottom, right) ©Calvin J. Hamilton; 20 (top) NASA; (center) EUV, Image Science Team, NASA; (bottom) DMR, COBE, NASA, Four-Year Sky Map; 21 (all) NASA; 22-23 NASA/JPL/University of Arizona; 24 NASA; 27 NASA/JPL; 28 (top) NASA/ESA, John Clarke (University of Michigan); (bottom) NASA/JPL; 28-29 NASA/JPL/University of Arizona; 30-31 (all) NASA/JPL; 32 (both) NASA/JPL; 33 (top) Imke de Pater/UC Berkeley/Lawrence Livermore National Laboratory; (bottom) NASA/JPL; 34-35 SOHO-LASCO Consortium, ESA, NASA; 37 H. Fukushima, D. Kinoshita and J. Watanabe (NAOJ); 38-39 (both) NASA/JHUAPL; 40 (all) NASA, H. Weaver and P. Feldman (JHU), M. A'Hearn (University of Maryland), C. Arpigny (Liege University), M. Combi (University of Michigan), M. Festou (Observatoire Midi-Pyrenees) and G.-P. Tozzi (Arcetri Observatory); 40-41 Arne Danielsen, Norway; 41 Greg Bacon (STScI/AVL); 42 SwRI; 42-43 NASA/JPL; 43 ©Calvin J. Hamilton; 44 (top) NASA/JPL; (bottom, left) NASA; (bottom, right) Courtesy of Pat Rawlings, Science Applications International; 45 (top) NASA; (bottom, both) Courtesy Alan Fitzsimmons, Queens University, Belfast; 46-47 SOHO-EIT Consortium, ESA, NASA; 48 Sebastien Gouthier; 50-51 (all) SOHO-EIT Consortium, ESA, NASA; 52 (both) TRACE; 53 NASA; 54 (top) TRACE; (bottom) M. Aschwanden et al. (LMSAL), TRACE, NASA; 55 SOHO-EIT Consortium, ESA, NASA; 56 NASA/JPL; 56-57 University of Minnesota; 58-59 European Southern Observatory; 63 NASA and The Hubble Heritage Team (STScI/AURA); 64-65 NASA and Ron Gilliland (STScI); 65 R. Williams (STScI), the HDF-South team and NASA; 66-67 ©Anglo-Australian Observatory/Royal Observatory, Edinburgh; 68 NASA, ESA, and The Hubble Heritage Team (STScI/AURA); 70 ©Anglo-Australian Observatory/Royal Observatory, Edinburgh; 71 NASA, A. Fruchter. and the ERO Team (STScI); 72-73 NASA, The Hubble Heritage Team (STScI/AURA); 74-75 (both) ©Anglo-Australian Observatory/Royal Observatory, Edinburgh; 76 ©Anglo-Australian Observatory; 77 NASA and The Hubble Heritage Team (STScI/AURA); 78-79 ©Anglo-Australian Observatory; 79 NASA, Donald Walter (South Carolina State University), Paul Scowen and Brian Moore (Arizona State University); 80-81 NASA and The Hubble Heritage Team (STScI/AURA); 85 FORS Team, 8.2-Meter VLT, ESO; 86 ©Jim Hurley; 87 (both) Eric Mamajek, University of Arizona Astronomy Department; 88-89 Christopher A. Davis/www.to-scorpio.com; 89 (all) D. Padgett (IPAC/Caltech), W. Brandner (IPAC), K. Stapelfeldt (JPL), NASA; 90 (top) C.R. O'Dell (Rice U.), NASA; (bottom) The National Science Foundation, NASA, and the Harvard-Smithsonian Center for Astrophysics; 91 Jeff Hester and Paul Scowen (Arizona State University), NASA; 92-93 ©Anglo-Australian Observatory; 94 G. Garmire et al. (PSU), NASA; 97 Dana Berry; 98-99 NASA and The Hubble Heritage Team (STScI/AURA); 99 Jeff Hester and Paul Scowen (Arizona State University), NASA; 100 NASA; 100-101 H. Bond (STScI), R. Ciardullo (PSU), WFPC2, HST, NASA; 102 Hubble Heritage Team (AURA/STScI/NASA); 103 (top) NASA; (bottom) Dr. Christopher Burrows, ESA/STScI, NASA; 104-105 Lynette Cook; 109 Lynette Cook; 110-111 NASA, UIT; 112-113 R. Williams and the HDF Team (STScI), NASA; 115 NASA; 116-117 ©Anglo-Australian Observatory; 118-119 NASA and The Hubble Heritage Team (STScI/AURA); 120 NASA/IOA/A. Fabian et al.; 121 Kirk Borne (Raytheon and NASA GSFC, Greenbelt, MD), Luis Colina (Instituto de Fisica de Cantabria, Spain), and Howard Bushouse and Ray Lucas (STScI, Baltimore, MD) and NASA; 122 NASA/JPL; 124-125 NASA/JPL/Caltech; 126 NASA/Ames Research Center, CA; 128 (top) NASA/JPL; (bottom) NASA; 129 (all) NASA/JPL; 130-131 MOLA Science Team; 132-133 NASA/JPL; 135 NASA/JPL Outer Planets/Solar Probe Project; 136 NASA/JPL; 137 (top) Courtesy Mark Showalter; (bottom) ©Calvin J. Hamilton; 138-139 JPL Picture Archive; 142 NASA; 143 (left & bottom) NASA; (right) ESA/D. Ducros; 144-145 TRACE; 146 SOHO-EIT Consortium, ESA, NASA; 148 TRACE; 149 (top) SOHO-EIT Consortium, ESA, NASA; (bottom) NASA; 150-151 TRACE; 152-153 NASA/JHUAPL; 154-155 (all) NASA/JPL; 156-157 NASA/JPL; 158 Department of Astrophysical and Planetary Sciences, University of Colorado; 159 NASA; 160-161 European Space Agency; 162 NASA; 165 NASA/Lockheed Martin Concept; 166-167 Naval Research Laboratory, DC, Astrophysics Branch; 168 (both) European Space Agency; 169 (both) NASA; 170-171 Smithsonian Institution and University of Arizona; 172 Mauna Kea Observatory, HI, Keck I & II; 174-175 European Southern Observatory; 176 FIRST survey image courtesy Robert H. Becker, Richard L. White and David J. Helfand; 176-177 European Southern Observatory; 178 (top) SDSS Collaboration; (bottom) Reidar Hahn, Fermilab; 179 (both) European Southern Observatory; 180 NASA; 182-183 NASA/JPL; 185 Planetary Society; 186 Courtesy W.M. Keck Observatory; 188-189 (both) NASA; 190-191 NASA; 192-193 Photo Researchers/Mehau Kulyk/Science Photo Library; 194 Benjamin C. Bromley/NASA/JPL/Caltech; 198 (top) Benjamin C. Bromley/NASA/JPL/Caltech; (bottom) NASA/HST; 199 Carol Osborne/CELT; 200-201 Painting by Dr. William K. Hartmann; 203 The Clementine Project; 204 Painting by Dr. William K. Hartmann; 206-207 (both) Photo Researchers/DetLev Van Ravenswaay/Science Photo Library; 208-209 Lynette Cook; 212-213 Gary J. Tonge, Vision Afar. Spaceart.org; 215 NASDA-Japan; 216-217 Gary J. Tonge, Vision Afar. Spaceart.org

SPACE ODYSSEY
Voyaging through the Cosmos

By William Harwood

PUBLISHED BY THE NATIONAL GEOGRAPHIC SOCIETY

John M. Fahey, Jr. *President and Chief Executive Officer*
Gilbert M. Grosvenor *Chairman of the Board*
Nina D. Hoffman *Executive Vice President*

PREPARED BY THE BOOK DIVISION

Kevin Mulroy *Vice President and Editor-in-Chief*
Charles Kogod *Illustrations Director*
Barbara A. Payne *Editorial Director*
Marianne Koszorus *Design Director*

STAFF FOR THIS BOOK

Tom Melham *Project Editor and Text Editor*
Marilyn Gibbons *Illustrations Editor*
Lyle Rosbotham *Art Director*
Winfield Swanson *Researcher*
Ron Fisher *Legend Writer*
R. Gary Colbert *Production Director*
Sharon Kocsis Berry *Illustrations Assistant*
Julia Marshall *Indexer*
Bey Wesley *Staff Assistant*

MANUFACTURING AND QUALITY CONTROL

George V. White *Director*
Vincent P. Ryan *Manager*
Phillip L. Schlosser *Financial Analyst*

Copyright © 2001 National Geographic Society.
All rights reserved. Reproduction of the whole or any
part of the contents without permission is prohibited.

Harwood, William, 1952-
   Space odyssey: voyaging through the cosmos / William Harwood
      p. cm.
   Includes index.
   ISBN 0-7922-6354-5 reg. -- ISBN 0-7922-6355-3 dlx.
   1. Solar system. 2. Cosmology

The world's largest nonprofit scientific and educational organization, the National Geographic Society was founded in 1888 "for the increase and diffusion of geographic knowledge." Since then it has supported scientific exploration and spread information to its more than eight million members worldwide.

The National Geographic Society educates and inspires millions every day through magazines, books, television programs, videos, maps and atlases, research grants, the National Geographic Bee, teacher workshops, and innovative classroom materials.

The Society is supported through membership dues, charitable gifts, and income from the sale of its educational products.

Members receive NATIONAL GEOGRAPHIC magazine—the Society's official journal—discounts on Society products, and other benefits.

For more information about the National Geographic Society, its educational programs, publications, or ways to support its work, please call 1-800-NGS-LINE (647-5463), or write to the following address:

National Geographic Society
1145 17th Street, N.W.
Washington, D.C. 20036-4688
U.S.A.

Visit the Society's Web site at
www.nationalgeographic.com

Printed in the U.S.A.